Self-Assessment Color Review of
Small Mammals

Susan A. Brown
DVM
Midwest Bird & Exotic Animal Hospital
Westchester, Illinois

Karen L. Rosenthal
MS, DVM, Dipl ABVP Avian
The Animal Medical Center
New York, New York

Iowa State University Press/Ames

Common name	Scientific name	Original range
13-lined ground squirrel	*Citellus tridecemlineatus*	North America
African hedgehog	*Erinaceus hindi*	Africa
Chinchilla	*Chinchilla laniger*	South America
Common marmoset	*Callithrix jacchus*	South America
Cotton-tail rabbit	*Sylvilagus nuttalii*	North America
Cotton-top tamarin	*Saguinus oedipus*	South America
Domestic rabbit	*Oryctolagus cuniculus*	Worldwide
Eastern gray squirrel	*Sciurus carolinensis*	North America
Eastern red squirrel	*Tamiasciurus hudsonicus*	North America
European hedgehog	*Erinaceus europaeus*	Europe
Ferret	*Mustela putorius furo*	Europe/North Africa
Golden hamster	*Mesocricetus auratus*	East Europe/Middle East
Guinea pig	*Cavia porcellus*	South America
Lion tamarin	*Leontopithecus rosalia*	Brazil
Long tailed macaque	*Maccaca fascicularis*	South-East Asia
Mongolian gerbil	*Meriones unguiculatus*	Mongolia
Mouse	*Mus musculus*	Worldwide
Owl monkey	*Aotus trivirgatus*	Central/South America
Patas monkey	*Erythrocebus patas*	Africa
Potbellied pig	*Sus scrota*	Asia
Prairie dog	*Cynomys ludovicianus*	North America
Raccoon	*Procyon lotor*	North America
Rat	*Rattus norvegicus*	Worldwide
Red fox	*Vulpes vulpes*	North America
Rhesus monkey	*Macaca mulatta*	India
Spider monkey	*Ateles geoffroyi*	South America
Squirrel monkey	*Saimiri sciureus*	South America
Stripped skunk	*Mephitis mephitis*	North America
Virginia opossum	*Didelphis virginiana*	North America

First published in the United States of America in 1997 by
Iowa State University Press, 2121 South State Avenue, Ames, Iowa 50014-8300
ISBN: 0-8138-2092-8

A CIP catalogue record for this book is available from the British Library.

Typeset and designed by: Paul Bennett
Colour reproduction: Reed Reprographics, Ipswich, UK
Printed by: Grafos SA, Barcelona, Spain.

Preface

Small mammal veterinary medicine has greatly expanded in its breadth of knowledge over the past ten years. During the same period, people owning pets such as rabbits, ferrets and rodents have demanded a higher quality of medical care than was formerly expected. These two factors have forced veterinarians to seek out sources of information about these species. Until recently, it has been difficult to obtain accurate, current, non-anecdotal written material on small mammals.

To aid veterinarians in their quest for information, we offer this text in a self-assessment format which challenges readers to use their clinical knowledge on a variety of species. We have assembled an international collection of distinguished colleagues that have contributed cases which range from basic to advanced. Therefore, this book can be used by veterinarians with a wide range of clinical experience.

The format is a question followed by its detailed answer on the next page. We have strived to construct every question and answer to 'stand alone' and not rely on information provided in other cases. The topics consist of case reports, clinical techniques, anatomy and husbandry. We have tailored the number of questions per species to reflect the frequency in which these species are seen in practice. As such, rabbits, ferrets and guinea pigs constitute the largest numbers of cases in the book.

A unique aspect of this book is that most questions are illustrated with a photograph, radiograph or diagram to aid in understanding a concept. Most case report answers include differential diagnoses to enable veterinarians to reason beyond the the scope of particular questions. Unlike many other small mammal texts, this is a very practical book with detailed solutions that allow veterinarians to put into use the information they have just read.

We anticipate that some readers will criticize the extent or complexity of the diagnostics and treatments suggested for some of the species in the text. Although financial constraints often determine the level of veterinary care patients receive, veterinarians are obliged to make owners aware of all the available options. We have offered in our answers the best possible treatment plans available at the time of writing and, as needed, other less complex or expensive options can be tailored from this information. We must, as veterinarians, offer the same high standard of care for these small mammal species as we do for other companion animals and let the owners decide for themselves the extent to which they wish to pursue treatment. The human-animal bond can extend to all species and it should not be dishonored by our prejudgments.

The text is not meant to be memorized. Small mammal medicine is a rapidly evolving field. The ultimate challenge to veterinarians is to apply the critical thinking this book demands, the fundamental veterinary medicine principles, and the basic knowledge of physiology, anatomy and husbandry to future situations that may be encountered in small mammal species.

Susan A. Brown
Karen L. Rosenthal
1997

Dedications

Richard, Zoki, Alex, Falcor, Uma, Travis, Lucy, Psycho Kitty, Cleo and Indy

Hilary, Eleyna, Bill and Tigger

Contributors

Nancy L. Anderson, DVM, Dipl ABVP Avian
Ohio State University
Columbus, Ohio USA

R. Avery Bennett, DVM, MS, Dipl ACVS
University of Florida
Gainesville, Florida USA

Beth Ann Breitweiser, DVM
All Wild Things Exotic Animal Hospital
Indianapolis, Indiana USA

Susan A. Brown, DVM
Midwest Bird & Exotic Animal Hospital
Westchester, Illinois USA

David A. Crossley, BVetMed, FAVD, MRCVS
Animal Medical Centre Referral Services
Chorlton, Manchester UK

Barbara J. Deeb, DVM, MS
University of Washington
Seattle, Washington USA

Pamela H. Eisele, DVM, Dipl ACLAM
University of California
Davis, California USA

Paul A. Flecknell, MA, VetMB, PhD, DLAS,
Dipl ECVA, MRCVS
University of Newcastle-upon-Tyne
Newcastle-upon-Tyne, Tyne and Wear UK

Karen Helton, DVM, Dipl ACVD
Cardiopet Dermatology Service
Little Falls, New Jersey USA

Elizabeth V. Hillyer, DVM
Oldwick, New Jersey USA

Jeffrey R. Jenkins, DVM, Dipl ABVP Avian
Avian & Exotic Animal Hospital
San Diego, California USA

Cathy A. Johnson-Delaney, DVM
University of Washington
Seattle, Washington USA

Douglas R. Mader, MS, DVM, Dipl ABVP
Long Beach Animal Hospital
Long Beach, California USA

Paul E. Miller, DVM, Dipl ACVO
University of Wisconsin
Madison, Wisconsin USA

Timothy H. Morris, BVetMed, PhD,
CertLAS, Dipl ACLAM, MRCVS
SmithKline Beecham Pharmaceuticals
Harlow, Essex UK

Holly S. Mullen, DVM, Dipl ACVS
Emergency Animal Hospital & Referral
Center
San Diego, California USA

Christopher J. Murphy, DVM, Dipl ACVO
University of Wisconsin
Madison, Wisconsin USA

Joanne Paul-Murphy, DVM, Dipl ACZM
University of Wisconsin
Madison, Wisconsin USA

Robert D. Ness, DVM
Midwest Bird & Animal Hospital
Westchester, Illinois USA

Barbara Oglesbee, DVM, Dipl ABVP Avian
Ohio State University
Columbus, Ohio USA

Sharon Patton, MS, PhD
University of Tennessee
Knoxville, Tennessee USA

Stuart L. Porter, VMD
Blue Ridge Community College
Weyers Cave, Virginia USA

Samuel M. Ristich, DVM
Indian Prairie Animal Hospital
Aurora, Illinois USA

Karen L. Rosenthal, MS, DVM,
Dipl ABVP Avian
The Animal Medical Center
New York, New York USA

Andrew Routh, BVSc, MRCVS
Stapeley Grange Wildlife Hospital & Cattery
Nantwich, Cheshire UK

David Scarff, BVetMed, CertSAD, MRCVS
Anglian Referrals
Norwich, Norfolk UK

Anthony J. Smith, DVM
El Paso Zoo
El Paso, Texas USA

Greg Whelan, BVMS, CertLAS, MRCVS
SmithKline Beecham Pharmaceuticals
Harlow, Essex UK

David L. Williams, MA, VetMB, PhD,
CertVOphthal, MRCVS
Animal Health Trust
Newmarket, Suffolk UK

1 This four-year-old African hedgehog died suddenly without premonitory signs (1).
a What are the lesion(s) observed on necropsy?
b What is the most likely cause of the postmortem changes in this animal?

2 A seven-year-old intact male rabbit is lethargic, cachexic and approximately 8% dehydrated. The mucous membranes are sticky and pale and the kidneys are smaller than average. The heart rate is 300 beats per minute and the respiratory rate is 28 breaths per minute. The rabbit has a history of chronic respiratory disease, which has been treated by the owner with an unknown dose of tetracycline in the water for the past two years. There has been a noticeable weight loss over the past two months and the rabbit has been anorectic for two weeks. The following are the serum biochemistry and hematology results:

Test	Patient value	Normal values
PCV	19%	33–50%
Plasma protein	74 g/l (7.4 g/dl)	54–83 g/l (5.4–8.3 g/dl)
WBC	7.2 × 10³/ml	5.2–12.5 × 10³/ml
CO_2	11.4 mmol/l (11.4 mEq/l)	16–38 mmol/l (16–38 mEq/l)
Calcium	3.0 mmol/l (12.0 mg/dl)	1.4–3.1 mmol/l (5.6–12.5 mg/dl)
Phosphorus	3.4 mmol/l (10.6 mg/dl)	1.3–2.2 mmol/l (4.0–6.9 mg/dl)
Glucose	5.6 mmol/l (100 mg/dl)	4.2–8.7 mmol/l (75–155 mg/dl)
Creatinine	689.5 μmol/l (7.8 mg/dl)	44.2–221 μmol/l (0.5–2.5 mg/dl)
BUN	191 mg/dl	13–26 mg/dl
Sodium	151 mmol/l (151 mEq/l)	131–155 mmol/l (131–155 mEq/l)
Potassium	6.4 mmol/l (6.4 mEq/l)	3.6–6.9 mmol/l (3.6–6.9 mEq/l)
Alanine aminotransferase	63 U/l	48–80 U/l
Urine specific gravity	1.010	1.003–1.036

a What is your diagnosis for the anorexia?
b What is your treatment regime for this rabbit?

3 Fentanyl/droperidol and fentanyl/fluanisone are useful for providing analgesia and immobilization for minor surgical procedures in rabbits. After completion of the procedure, some of the effects of these neuroleptanalgesic mixtures are reversed with which one of these drugs: butorphanol, naloxone, atipamezole or yohimbine?

1–3: Answers

1 a The most prominent gross lesion is splenomegaly. Hepatomegaly is also present.
b Neoplasia. Numerous types of neoplasia are described in all hedgehog species including primary or metastatic adenocarcinoma, hemangiosarcoma, lymphoma, fibrosarcoma, hepatoma, hepatocellular carcinoma and leiomyosarcoma. Animals over three years of age are most susceptible. In one study, 69% of individuals that died in an African hedgehog collection had evidence of an ongoing neoplastic process. Other differential diagnoses for splenomegaly and hepatomegaly include septicemia, trauma and cardiac disease.

2 a The presence of markedly elevated BUN and creatinine values in the face of isosthenuric (not concentrated) urine is diagnostic for renal failure. Chronic renal failure is more likely than acute renal failure in this rabbit because of the presence of small kidneys, hyperphosphatemia and anemia, and the lack of hyperkalemia and oliguria. Causes for acquired chronic renal failure in pet rabbits have not been studied. Common etiologies cited for other mammals include infectious interstitial nephritis or pyelonephritis (bacterial, viral, fungal), glomerulonephritis secondary to immune complex deposition, neoplasia and nephrotoxins, such as antibiotics and hypercalcemia. Especially in consideration of the history of chronic respiratory infections, obtain a urine culture to rule out bacterial causes of the renal failure. The effects of the chronic administration of tetracyclines in the drinking water on the renal failure is unclear. Tetracyclines can exacerbate chronic renal disease and overdosing is associated with acute renal failure in other species.
b Administer intravenous or intraosseous lactated Ringer's solution immediately. Replace the dehydration deficit over six hours and continue with diuresis. Place the patient on standard maintenance fluids of 100 ml/kg/day once BUN concentrations approach more normal values. Substitute subcutaneous fluids once the dehydration is corrected and the patient is urinating normally. Monitor serum potassium and correct for any excesses or deficiencies. Much of the acidosis is expected to be corrected by the fluid therapy. Treat the anemia with ferrous sulfate (4–6 mg/kg PO q 24 hours). Add B-complex vitamins to the parenteral fluids if needed. Although dosages are not established for rabbits, anabolic steroid or recombinant human erythropoietin therapy may be beneficial. Treat the anorexia with gavage or tube feedings until the patient is eating on its own. Select a safe broad-spectrum antibiotic with minimal nephrotoxicity while awaiting culture results. Enrofloxacin (10–15 mg/kg SC, IM or PO q 12 hours) is an excellent first choice. Avoid trimethoprim/sulfa combinations, tetracyclines and aminoglycosides because of potential nephrotoxicity. Do not use chloramphenicol if anemia is present. If the rabbit shows minimal response to fluid therapy, the prognosis for long-term survival is grave.

3 Use naloxone or butorphanol to reverse the fentanyl component in the neuroleptanalgesic mixtures. These two drugs reverse the respiratory depression, reduce the degree of sedation and generally speed recovery. It is important to note that naloxone, a mu-opioid antagonist, also reverses the analgesic effect of the fentanyl, so that the rabbit may then experience pain. Administration of butorphanol reverses the effects of fentanyl but butorphanol has its own analgesic effect at K-opioid receptor(s), thus maintaining some postprocedure pain relief. Butorphanol is, therefore, preferred.

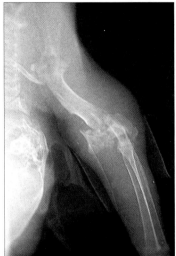

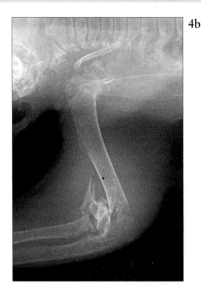

4 The radiographs demonstrate a 'Y' fracture of the distal humerus in a 178 g Eastern red squirrel (**4a, b**). External coaptation was elected for fracture stabilization due to the small size of the fracture fragments. A modified Schroeder–Thomas splint, using a wire coat hanger, is applied to the limb. This type of splint creates traction at the fracture site while maintaining the limb in a functional position. How would you create traction at the fracture site?

5 A domestic ferret has a systemic illness and you need to obtain a blood sample in order to diagnose disease.
a From where and how would you collect blood from the ferret?
b What volume of blood can be safely collected from a healthy ferret?
c Is blood typing necessary for a ferret blood transfusion?
d What sites are suitable for intravenous or intraosseous catheterization?

4 Place tape around the caudal aspect of the proximal humerus and then wrap it around the cranial bar of the splint. This secures the proximal humerus to the cranial bar of the splint. Then, secure the proximal portion of the antebrachium to the caudal bar of the splint by passing tape cranial to the antebrachium and around the caudal bar of the splint. By pulling the proximal radius and ulna caudally and the proximal humerus cranially, traction is applied to the fracture site. Apply additional tape support around the entire splint after the traction straps are applied to provide lateral and medial support.

5 a Cephalic vein, jugular vein, lateral saphenous vein, and ventral tail artery. Use a 25–27 gauge needle for small-diameter veins and a 22 gauge needle for the jugular vein and ventral tail artery. The novice practitioner may find it easier to perform phlebotomy in the anesthetized or sedated ferret. Most ferrets can be phlebotomized fully awake with the aid of a trained handler once some experience is gained.

Obtain blood from the cephalic vein by restraining the ferret as in the dog, with one hand around the neck and the other holding off the vein with the thumb with the forearm restraining the patient's body (illustrated). Once blood is seen in the hub of the needle, release the pressure with the thumb to allow more blood to flow into this short vein. About 0.5–1.0 ml of blood can be collected from this site. To access the jugular vein, have an assistant scruff the ferret and then wrap its body tightly with a towel restraining the ferret's forearms against its chest. Have the assistant maintain a scruff hold on the upper neck area and lay the ferret on its dorsum. Apply pressure just lateral to the thoracic inlet and visualize or palpate the jugular vein which runs from the inlet to the base of the ear. Clip the hair if needed to aid in visualization. The jugular vein can also be accessed by holding the ferret in a similar manner as the cat with its front legs pulled over the edge of the table and the neck hyperextended.

Access the ventral tail artery by wrapping the ferret in the same manner as described for jugular venipuncture. Have the assistant scruff the patient and pull the hind legs cranially. Insert a 22 gauge needle in the ventral tail midline approximately 2.5 cm from the tail base and move the needle dorsally creating gentle suction until blood enters the syringe. Up to 3 ml of blood can be obtained from this site.

The lateral saphenous is a short vein that yields up to 0.5 ml of blood. Access this vein by holding the ferret on its side with a scruff on the neck and firm pressure applied to the upper thigh.
b A maximum of 1% of the body weight can be removed as blood volume in a healthy ferret. For example, a 1,000 g ferret can lose a maximum of 10 ml of blood safely.
c Ferrets have no discernable blood types, therefore blood typing is not necessary and multiple donors may be used to transfuse a ferret patient.
d The cephalic, lateral saphenous and jugular veins. The most common site for an intraosseous catheter is the femur. Approach through the trochanteric fossa and use a 20 gauge spinal or regular hypodermic needle.

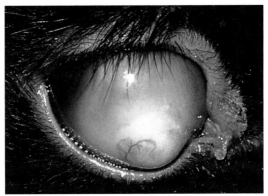

6 This three-year-old spayed female rabbit has a progressively enlarging white area in its eye (6). Currently, there is a bulge in the cornea over the white area.
a What is your diagnosis and what is the likely etiologic agent?
b What treatment would you recommend?
c Why is this a challenging procedure in rabbits?
d Describe two enucleation techniques in rabbits.

7 A three-year-old female guinea pig has had an unpleasant body odor for three days. Six days ago the owner took the guinea pig to another veterinarian to have the feet examined because the plantar surfaces were reddened (7). The veterinarian diagnosed pododermatitis. The owner was instructed to take the guinea pig off wire flooring and place it on shredded paper. The veterinarian also prescribed oral amoxicillin with clavulinic acid to treat the infection. The feces in the carrier have a thick-liquid consistency. The guinea pig is still eating and drinking.
a What is the source of the odor?
b What is the cause of the soft stool?

6, 7: Answers

6 a An intraocular abscess most likely due to *Pasteurella multocida*.
b Enucleation.
c The very large orbital venous sinus can cause significant, even fatal hemorrhage, if it is lacerated.
d Two enucleation techniques in rabbits are described that avoid cutting the sinus. Indications for enucleation include intraocular *Pasteurella* spp. abscesses, trauma and other end-stage ocular diseases. The first technique is a transconjunctival approach. Remove the borders of the eyelids and dissect the palpebral conjunctiva off the surface of the eyelids. Continue dissection on to the bulbar conjunctiva along the globe. Stay as close as possible to the surface of the globe to avoid damaging the venous sinus. Once all the periorbital structures are dissected free from the globe, place a hemostatic clip on the optic nerve and blood vessels and remove the globe.
   Alternatively, use a transpalpebral approach. Perform dissection staying as close to the wall of the bony orbit as possible in order to avoid damaging the venous sinus. Remove all structures from within the bony orbit. If the sinus begins to bleed, place hemostatic clips blindly in order to control the hemorrhage. As a last approach, remove the globe as quickly as possible, pack the bony orbit with gel foam, and apply digital pressure to control hemorrhage. Remove all conjunctival and glandular tissue prior to closure. Suture the eyelids closed once hemostasis is achieved. In most rabbits, 3-0 or 4-0 nylon works well.

7 a Feces soiled perineal area. Other causes of strong body odor in the guinea pig include perineal soiling due to urinary tract or uterine disease, soiling of fur around the mouth due to dental disease, smegma in the perineal sac (which occurs in both male and female) and infections or neoplasia of the perianal sebaceous glands.
b Antibiotic associated diarrhea of guinea pigs. The use of Gram-positive spectrum oral antibiotics can seriously alter the sensitive balance of bacterial flora in the cecum. These antibiotics decrease the populations of dominant bacteria allowing species such as *Clostridia* spp. and *Escherichia coli*, which are normally present in small numbers, to colonize the cecum rapidly. These bacteria produce enterotoxins which are absorbed systemically. The result of the cecal dysbiosis and enterotoxin production is diarrhea, systemic toxicosis and ultimately death. Remember antibiotics to be avoided by the 'PLACE' rule (penicillin, lincomycin, ampicillin, amoxicillin, cephalosporins, clindamycin, erythromycin). Sulfa-trimethoprim combinations, enrofloxacin and chloramphenicol are rarely associated with GI disturbance in guinea pigs.

8 How would you treat the guinea pig in question 7?

9

9 List the precautions the person in the photograph should take when examining this wild skunk (9)?

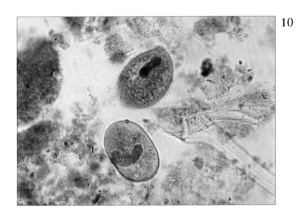

10

10 A recently purchased guinea pig develops diarrhea. On a direct fecal smear oval organisms measuring 55–115 × 45–75 μm are seen swimming across the slide (10). Similar organisms are found on a sugar flotation of the feces.
a What is the organism shown?
b Is this organism the probable cause of the diarrhea?

8 Conservative therapy can be used in this case because the guinea pig is still eating, drinking and is active and alert. Discontinue the antibiotic immediately. Offer a diet that consists primarily of free choice grass hay, which will aid in restoring normal gut motility. In addition, remove all grain-based or other high-starch treat foods and limit the pelleted foods. High levels of ingested starch encourage the growth of *Clostridium* spp. and *E. coli* in the cecum. Administer vitamin C (30–50 mg/kg PO q 12–24 hours) using either a syrup or crushed tablet. Vitamin C is a dietary requirement in the guinea pig and in cases of disease a dose higher than normal maintenance can be beneficial. In addition, vitamin C may slow the absorption of enterotoxins through the cecal wall. Use fresh dark, leafy greens, such as kale, mustard greens, dandelion greens, Swiss chard or collard greens, at half to one cup daily for the vitamin C source when normal stools are produced. Once greens are being consumed, discontinue oral dosing of vitamin C. Do not depend on pelleted foods to provide the complete vitamin C requirement due to this vitamin's relatively short shelf-life.

Check fecal Gram stains for clostridial spores if the diarrhea persists or if the pet becomes systemically ill. Treat with metronidazole (20 mg/kg PO q 12 hours) if spores are present. Obtain a fecal culture if spores are not visualized. Treat the guinea pig with injectable aminoglycosides or enrofloxacin while awaiting culture results. Use intravenous, subcutaneous or intraosseous fluids as needed in the systemically ill animal. The prognosis is grave in guinea pigs that are systemically ill with profuse watery diarrhea.

9 (1) Protect yourself and your staff from being bitten. Skunks can be asymptomatic carriers of rabies without outward signs of the disease. If a bite occurs, euthanize the animal and have the brain examined for rabies. Advise clients of the risk of rabies even in captive-raised animals. Consider having the professional staff vaccinated for rabies if exposure to this species is frequent. (2) Be aware of the legal obligations regarding the handling and treating of skunks. In some areas of the USA, it is forbidden to own or treat skunks. If a state requires a special permit to own this species, place a copy of this legal document in the client's record. (3) Vaccinate appropriately. Skunks need to be vaccinated for canine distemper only and can receive a modified-live distemper vaccine approved for use in ferrets. Vaccines seropassed in ferrets may result in clinical disease if given to another mustelid such as the skunk. There is no approved rabies vaccine for use in skunks. In addition, it is strictly forbidden to administer a rabies vaccine to wildlife species in most areas of the USA. (4) Be familiar with the husbandry requirements and the behavioral aspects of this species before accepting them as patients.

10 a This is a trophozoite of a *Balantidium* species, probably *B. caviae*. This is a ciliated protozoan and both cysts and trophozoites are found in the feces. It is identified by its large size and the swimming movements of the trophozoite. A characteristic macronucleus is visible in both the cyst and the trophozoite.
b This organism is usually non-pathogenic and is unlikely to be the primary cause of the diarrhea. Other causes for the diarrhea need to be investigated.

11 An African hedgehog has rolled tightly into a ball and has stayed in this position for some time (**11**).
a How would you collect a blood sample from this hedgehog?
b How would you administer parental fluid?

12 A mouse is maintained on gas anesthesia (**12**).
a What is an appropriate fresh gas flow rate for a mouse on an open-mask system?
b Does providing this flow rate present any special problems?

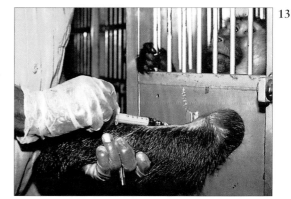

13 It is common to perform venipuncture on non-human primates (**13**).
a What venipuncture site would you recommend for obtaining large volumes of blood from an anesthetized primate?
b What venipuncture site would you recommend for the awake primate?

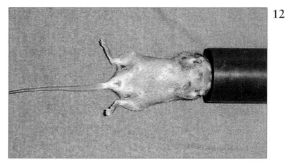

11 a Except in the very debilitated animal, it is usually necessary to anesthetize a hedgehog to obtain a blood sample. Isoflurane anesthesia is preferred for short procedures, although injectable anesthetics can also be used.

Visualize the superficially located lateral or medial saphenous vein. Collect blood either by lancing the vein with a needle and collecting the free flow of blood or by drawing blood directly into a syringe. Use small-gauge needles (25–27 gauge). Insulin syringes with the needle attached work well. The femoral vein is also a good collection site. Palpate this vein on the medial aspect of the thigh as pressure is applied at its proximal end. The delicate cephalic vein collapses easily and is a poor choice for venipuncture. The jugular vein is very short and lies deep beneath subcutaneous fat in the hedgehog, making it difficult to palpate. It runs from the thoracic inlet to the base of the ear. Apply pressure just lateral to the thoracic inlet on the preferred side of venipuncture. Do not use toe nail clipping as a routine method of blood collection in the hedgehog. It is painful and yields a very small, potentially contaminated, sample.

b Intravenous catheterization is difficult in the hedgehog, therefore the intraosseous route is used. Place an intraosseous catheter in the femur with the approach through the trochanteric fossa using a 20–22 gauge hypodermic needle of appropriate length. The catheter can be left in place for several days. Administer a systemic antibiotic while the catheter is in place and for at least three days afterwards.

12 a If the tidal volume of most animals is approximately 10–15 ml/kg then each breath of a 30 g mouse will be about 0.3–0.5 ml. Prevent rebreathing of exhaled gases in an open-mask system by using a flow rate of three times the animal's minute volume. Since respiratory rates in anaesthetized mice are generally 80–120 breaths/minute, this requires a flow rate of 75–180 ml/minute.

b Although this represents the most economical use of anesthetic gases, some vaporizers are incapable of delivering accurate vapour concentration at such low settings. Therefore, since it may be impractical to provide such a low flow, use a flow of 1–2 l/minute.

13 a The femoral vein. Locate the femoral vein in the femoral triangle. The pulse of the femoral artery helps to locate the triangle. The inguinal lymph node is often palpable in the cranial portion of the femoral triangle. Aim the needle and syringe in the direction of the pulse of the femoral artery. If the angle is correct, the vein will be entered first before the artery. Another common site to use is the saphenous vein on the posterior aspect of the hind limbs along the tibia. It courses along the posterior midline and is easily accessible even in small monkeys.

b The cephalic vein lies along the antebrachium. Primates can be trained to present their forearm through a cage opening to allow venipuncture.

14 A five-month-old female guinea pig has patchy hair loss and pruritus. Microscopic examination of a skin scraping reveals this organism (14).
a What is the genus and species of the organism?
b What treatment would you recommend?

15 Rabbit endotracheal intubation is challenging.
a What factors would you consider when endotracheal intubation is performed in a rabbit?
b What equipment would you use to perform endotracheal intubation in a rabbit?

16 A five-year-old pet rabbit is losing weight and having difficulty with prehension and chewing. Signs were first noticed several months ago and the rabbit's condition has been deteriorating. The rabbit is emaciated and has an abnormality of the incisor dentition (16) and a palpable irregularity of the ventral border of the mandible. What is the likely cause of the abnormality and how would you manage it?

14 a This parasite is *Gliricolla porcelli*, a debris-feeding louse. Lice and eggs attach to hair shafts and can be seen on gross inspection.
b Treat with ivermectin (0.2–0.4 mg/kg) repeated at two to three week intervals until the infestation is eliminated. This is an effective treatment because the lice ingest cutaneous fluids and therefore the medication. Other suggested treatments include insecticidal shampoos, dips, sprays or powders that are safe for kittens, repeated at the same frequency as the ivermectin. Use topical insecticides with extreme caution, particularly in debilitated or overweight animals. Treat all guinea pigs in a colony and clean the environment thoroughly.

15 a The rabbit has a long, narrow mouth that, particularly in smaller breeds, cannot open wide enough to permit simultaneous positioning of the laryngoscope blade, visualization of the larynx and placement of an endotracheal tube. Lightly anesthetized rabbits usually have very active laryngeal reflexes that contribute to the difficulties of placing an endotracheal tube. It is easy to traumatize the delicate oral, pharyngeal and airway tissues of rabbits with the laryngoscope blade or endotracheal tube stylet. Bleeding into the back of the mouth can lead to aspiration of blood. Traumatic intubation can also lead to cellulitis in the cervical region. Trauma to the trachea can cause soft-tissue swelling, leading to dyspnea after the tube is removed. An anesthetic mask, used for animals not intubated, does not allow for control of the airway in an emergency or for the use of mechanical ventilation.
b Endotracheal intubation equipment depends upon the practitioner's preference, the size of the rabbit and the method of intubation. Use Cole, straight or cuffed endotracheal tubes, 2.0–4.0 mm ID in size. For rabbits under 2 kg, use small laryngoscope blade sizes 58 mm or 75 mm and the larger blade size 102 mm for rabbits above 2 kg. Use either a Miller, Wisconsin or Guedel blade. Direct the placement of the endotracheal tube with commercially available semi-rigid endotracheal tube stylets or 3.5 or 5.0 Fr urinary catheters.

16 Obliquely worn incisor teeth may be due to functional disturbances, occlusal problems or uneven tooth growth. The most likely cause of this degree of oblique wear is uneven tooth regrowth following trauma. The rabbit was dropped approximately one year previously, fracturing both upper first incisor and left lower incisor teeth above the gingiva. The right lower incisor was not trimmed until the fractured teeth had regrown resulting in abnormal occlusion. The right upper first incisor was also growing back at an abnormal angle, thus exacerbating the situation. In cases showing a mild degree of oblique incisor wear, an abnormal chewing action, such as that caused by a unilateral cheek tooth problem, should be suspected. To prevent this problem, trim the incisors every two to four weeks to allow proper tooth alignment as they regrow. In addition, perform a complete physical examination, including visualization of the molars, with each visit.

In cases where the incisors fail to grow back into occlusion, extraction is a viable option provided that the cheek teeth are healthy. The alternative is regular trimming of the incisor teeth, preferably using a dental burr, at four to six week intervals for the remainder of the life of the rabbit.

17 a With regard to the rabbit in question **16**, what additional cause would you suspect, particularly if this was a 15-month-old dwarf breed rabbit?
b Why would you perform further investigations prior to attempting dental treatment?

18 Ketamine is a commonly used anesthetic. Why is it not recommended to administer intramuscular ketamine to small rodents?

9a

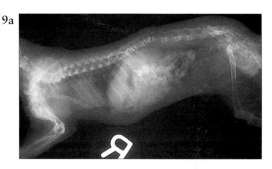

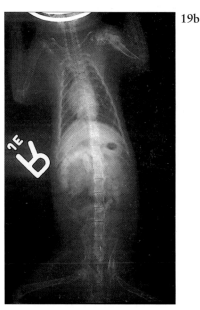

19b

19 A juvenile male squirrel, cared for by a wildlife rehabilitator, is no longer able to walk. There is pain and swelling in the left shoulder. The squirrel's diet consists of only fruits and nuts. These radiographs are taken (**19a, b**).
a What is your diagnosis?
b What is the name of the metabolic process responsible for this problem and how did it develop?
c How would you treat the orthopedic problems?

17

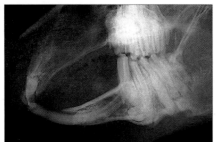

17a Dwarf breed rabbits are particularly prone to hereditary malocclusions including relative mandibular prognathism. These usually become apparent by 12 months of age, causing serious dysfunction by the time the animal is 18 months old. Tooth overgrowth occurs, impeding mastication, which ultimately leads to starvation if not treated. Unfortunately, affected individuals may have already produced several litters by the time a problem is recognized.

b Secondary cheek tooth malocclusion is common after a prolonged period of incisor dysfunction. The reverse is also true. Once a significant degree of secondary malocclusion develops, repeated dental procedures are needed for maintenance. Perform a thorough examination of the mouth along with radiography of the skull while the patient is anesthetized. This provides the best chance of detecting both intra-oral cheek tooth problems and tooth root disease. In this case, the maxillary and mandibular incisor tooth roots are abnormal following the earlier trauma. The intra-oral overgrowth of the first mandibular premolar teeth shows as a clearly stepped occlusal line on the lateral radiograph (17). Grossly deformed mandibular cheek tooth roots are also visible. Palpable irregularities of the ventral mandible are common in cases of cheek tooth malocclusion in rabbits due to a high incidence of root elongation, and are an indicator of a poor long-term prognosis.

18 IM ketamine causes muscle necrosis and the volume of injection required is too high and will be painful. The low pH of ketamine (3.0) also results in pain on injection. The relatively high doses required to produce sedation (50–100 mg/kg) results in injecting a relatively large volume of drug in comparison to other species. This larger relative volume may explain the increased occurrence of muscle necrosis.

19 a Compression fractures of the proximal left humerus and of the second lumbar vertebra are seen radiographically. There is also a general decrease in cortical thickness in all long bones.

b Nutritional secondary hyperparathyroidism is the metabolic process responsible for the bony disease. Fruits and nuts are low in calcium and relatively high in phosphorus. The result of a diet with a reverse calcium:phosphorus (Ca:P) ratio is metabolic bone disease. The compression fractures are pathologic.

c The overall goal of treatment is to place the animal on a proper plane of nutrition by correcting the diet, specifically the Ca:P ratio. Administer parenteral calcium once and follow with oral supplementation of calcium. Correct the diet. No stabilization is recommended for the fractures. These bones are too soft to hold implants. Pins often tear through and wires collapse the bone. Confine the squirrel to a small cage and handle with great care. Generally, pathologic fractures due to nutritional secondary hyperparathyroidism heal rapidly once the diet is corrected. Remodeling of malunion fractures generally results in a functional outcome. In a compression vertebral fracture, if pain perception is present, the prognosis is fair to good for a complete recovery. It must be emphasized that gentle handling and restricted activity is vital, as well as an improved diet.

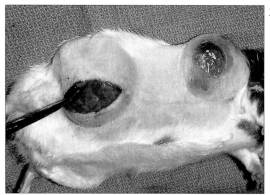

20 A two-year-old female intact pet rat develops these two large tumors on the abdomen (**20**).
a What is your diagnosis and prognosis?
b How does this condition in rats compare with that in mice and guinea pigs?
c What would you recommend to prevent the occurrence of these tumors in pet rats?

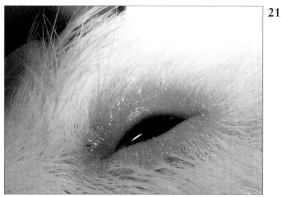

21 A two-year-old guinea pig is depressed, thin and anorexic with a clear, watery discharge from both eyes (**21**). The chin is wet from saliva and there is a serous discharge and crusting around the nose.
a What considerations must you take into account when performing a physical examination on an ill guinea pig?
b What is the most likely cause of the guinea pig's condition?
c What is the cause of the oculonasal discharge?

20, 21: Answers

20 a This is most likely a mammary gland tumor. Mammary tumors are common in rodents. In rats, these tumors are usually benign mammary fibroadenomas. Although, metastasis is rare, perform radiographs before surgery. Tumors generally metastasize to regional lymph nodes, abdominal viscera or lungs. During surgical removal of mammary tumors, collect regional lymph nodes to stage the disease. Unless ulcerated or suspected to be malignant, do not remove the skin overlying the tumor. Ligate any blood vessels supplying the tumor. Closure is routine. Use a subcuticular suture pattern or skin staples because rodents frequently chew and may remove the sutures. Use an Elizabethan collar as needed. Prognosis for recovery is good unless there is metastasis.
b In mice, mammary tumors are frequently malignant, invasive and difficult to remove. They are associated with infection by the mouse mammary tumor virus. In guinea pigs, about 70% of the mammary gland tumors are benign fibroadenomas and 30% are mammary adenocarcinomas.
c Ovariohysterectomy at a young age may decrease the incidence of mammary gland tumors and can be recommended to owners of female pet rats.

21 a Guinea pigs have a relatively low tolerance for pain. Inappropriate handling of an ill guinea pig, particularly if it is in pain, can be disastrous. Warn owners prior to an examination that handling severely ill patients can be risky. If the patient is depressed and uncomfortable, perform the examination as quickly and gently as possible and have oxygen readily available. Do not place an ill guinea pig in dorsal recumbency particularly if respiratory or cardiac disease is suspected. If necessary, use inhalant anesthesia to reduce the stress of examination or diagnostic sampling. Use pain relieving medications, such as butorphenol, buprenorphine and flunixin meglumine, as needed.
b The most common cause of anorexia coupled with a wet chin in a guinea pig is overgrown molars. Many factors influence the severity of this condition (see 87).
c The stress of handling can cause clear to milky lacrimal and nasal secretions in the guinea pig. Upper respiratory disease and pneumonia caused by *Bordetella* and *Klebsiella* spp. or viral agents can also result in oculonasal discharge. *Chlamydia psittaci* is a cause of guinea pig inclusion conjunctivitis. Other causes of oculonasal discharges include chronic hypovitaminosis C and exposure to noxious odors, such as ammonia or volatile oils, from pine or cedar shavings.

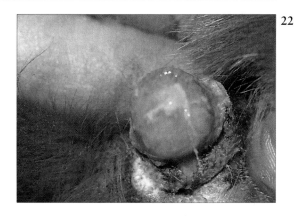

22 A five-year-old intact male guinea pig is thin, painful on abdominal palpation and had a decreased appetite for one month. There is a urethral plug visible (22).
a What are the two most likely etiologies for the plug?
b What is the best diagnostic test to differentiate the etiologies?
c If the result of the test is negative, what other diagnostic tests would you perform?
d How would you treat it while you await the laboratory results?
e What long-term husbandry would you recommend?

23 A rabbit is induced with gas anesthesia (23). What is a potential problem during induction of anesthesia in rabbits with volatile agents, such as isoflurane or halothane?

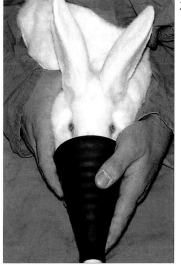

22 a Inspissated secretions from the vesicular glands or urethral calculi.

b An abdominal radiograph. Most calculi in guinea pigs are formed from calcium or magnesium salts and are therefore radiopaque. Calculi may also appear in the ureters. Concretions in the vesicular glands can also appear radiopaque, but will be located outside the areas of the ureters or bladder.

c Obtain a urinalysis and urine culture by cystocentesis to evaluate for an infectious cystitis. Examine a serum biochemistry profile to evaluate renal function secondary to the obstruction.

d Relieve the obstruction using a water-soluble lubricating jelly and gentle digital pressure. A normal stream of urine should be observed as the bladder is gently expressed. If urethral patency is questionable, sedate the guinea pig with isoflurane and pass a small, soft catheter into the urethra and flush with sterile saline. Analgesics may be necessary post-catheterization to prevent urethral spasm.

In cases where the obstruction cannot be relieved quickly, perform a cystocentesis with a 25–27 gauge needle to relieve the pressure in the bladder. If the urethra remains obstructed, perform a cystotomy to allow retrograde flushing and removal of the plug. Administer fluid therapy with 0.9% saline using an intravenous or intra-osseous catheter in obstructed guinea pigs that are systemically ill. In stable individuals, use 0.9% saline subcutaneous to support renal function. Many obstructions are associated with a bacterial infection. Administer enrofloxacin while awaiting culture results. Use trimethoprim/sulfa combinations only if minimal renal damage is suspected. Administer oral vitamin C (30–50 mg/kg IM, SC or PO q 12–24 hours) to aid in the healing of damaged tissue, particularly in anorectic animals. Vitamin C is a necessary dietary requirement of guinea pigs.

e Gently clean secretions from the prepuce and perineal area as needed with a mild antiseptic solution once a week. Put the guinea pig on a healthy diet that contains free choice grass hay, limited pellets and a minimum of a half cup daily of dark leafy greens to provide essential dietary vitamin C. Vitamin C can be added to the drinking water, but it has a bitter taste and is inactivated in less than 24 hours. A bitter taste in the water may decrease water consumption, which is not desirable in an animal with potential renal disease.

23 Breath-holding is a potential problem in rabbits. Immediately following exposure to halothane or isoflurane, some rabbits hold their breath for periods of one to two minutes. Often this is not noted because of the close physical restraint employed to prevent the animal from injuring itself due to struggling. Breath-holding results in a reflex bradycardia and also moderate hypercapnia and hypoxia. Since breath-holding is a voluntary response, it is reasonable to assume that the rabbit is stressed by the procedure and its plasma catecholamine concentration is elevated. When the rabbit does eventually inhale, it may be exposed to a high concentration of halothane (which is known to sensitize the heart to the arrhythmogenic effects of catecho-lamines) and may also be hypoxic and hypercapnic. This may explain the sudden death of some rabbits on induction. Unfortunately, the use of commonly available pre-anesthetic medications (e.g. acepromazine, xylazine) does not eliminate the problem. Eliminate this problem by observing the animal closely and remove the mask when breath-holding occurs. Reintroduce the mask as the rabbit recommences breathing.

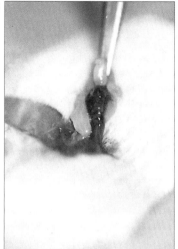

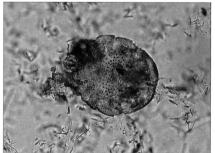

24 This photograph demonstrates a transverse, dorsal approach for bilateral ovariectomy in mice (24). How would you carry out the technique?

25 A four-year-old ferret has this skin mass on its foot (25).
a What are your differential diagnoses for ferret skin masses?
b What is the most likely diagnosis?
c How would you treat the masses?
d What is the prognosis?

26 A mouse develops pruritus and hair loss over its dorsum. The skin appears reddened with multifocal crusts and patchy alopecia. Another mouse in the same cage is asymptomatic. There is also a 10-year-old Yorkshire terrier recently diagnosed with allergic dermatitis living in the same household.
a What are the differential diagnoses for pruritus and alopecia in the mouse?
b What diagnostic steps would you take to find the cause of the problem?
c What is the organism shown and how would you treat it (26)?
d What recommendations would you make regarding the other animals in the household?

24 In small rodents it is possible to remove the ovary (without removing the uterus) through a lumbar approach. Place the patient in ventral recumbency and prepare the lumbar and flank areas for aseptic surgery. Make a dorsal midline incision or, alternatively, make a transverse incision on each side of the dorsal midline. Through the incision, shift the skin from one side to the other to gain access to each ovary. Through the skin incision, bluntly dissect caudal to the last rib at approximately the level of the third lumbar vertebra through the muscles of the body wall into the peritoneal cavity. The ovary is located within the fat pad at the caudal pole of the kidney. Remove the ovary. No ligation is required and hemorrhage is generally minimal although a hemaclip can be placed on the vessels, if needed. Appose the muscle with 5–0 or 6–0 synthetic, absorbable suture material and close the skin with subcuticular sutures.

25 a They include mast cell tumors, sebaceous gland adenomas, squamous cell tumors, adenocarcinomas, cutaneous lymphoma and abscesses.
b Mast cell tumors are the most frequently seen cutaneous neoplasia in the ferret. Cutaneous mast cell tumors in ferrets appear as papules or nodules and vary in color from tan, yellow, brown or red. They are multiple or solitary and occasionally will resolve without surgical removal. Some are pruritic and are covered with dried blood. Lesions can occur anywhere. Mast cell tumors in the ferret are usually benign but metastasis to distant organs has been reported.
c Excise and biopsy these masses under general anesthesia. Pretreatment with antihistamines is not indicated. Radiograph and/or ultrasound ferrets with skin masses to examine for the presence of organomegaly due to distant neoplasia.
d The long-term prognosis for the majority of cases of mast cell tumor is good.

26 a The causes of pruritus and alopecia in the mouse include ectoparasites, dermatophytosis, bacterial infections, barbering, pinworms and neoplasia. The more common ectoparasites include the parasitic mites *Myobia musculi* and *Myocoptes musculinus*. Less common species of mites include *Radfordia affinis*, *Psorergates simplex*, *Otodectes bacoti*, *Sarcoptes scabiei*, *Notoedres muris* and *Trichoecius romboutsi*. Other ectoparasites include fleas, primarily in households with dogs and cats. The louse, *Polyplax serrata*, is also considered. Dermatophytosis is an uncommon cause of pruritus in the mouse. *Trichophyton mentagrophytes* can be isolated from 60% of clinically normal pet shop mice. Bacterial dermatitis (i.e. *Staphylococcus*) is not typically pruritic but may cause skin crusting. Barbering, due to stress and overcrowding, causes alopecia but not pruritus. Pinworms cause a localized perianal pruritus. Neoplasia, such as epitheliotrophic lymphoma, occasionally causes alopecia and pruritus but generally is associated with scaly skin.
b Perform skin scrapings, fungal cultures and skin biopsies to determine the cause.
c This *Sarcoptes* mite was found after a skin scraping. Two to three injections of ivermectin (0.20–0.40 mg/kg q 7–14 days) is usually sufficient to cause resolution of the signs. Insecticidal dips may be too toxic, even diluted, for mice. Use any insecticide product in mice with great caution.
d Consider the other mouse an asymptomatic carrier and treat it with ivermectin. Since the dog may not be suffering from allergies but rather a mite infestation, it is important to examine all household pets and people for mites. The household environment should be thoroughly cleaned.

27 Chinchillas have only recently become popular as pets (27).
a What is the origin of the domesticated chinchilla?
b What are the special husbandry requirements of chinchillas?
c Describe the anatomical characteristics of the male and female chinchilla.

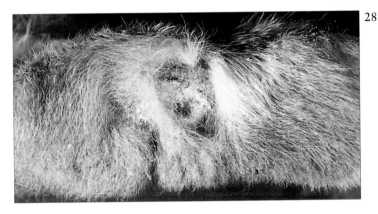

28 A four-year-old obese female rabbit develops ulcerations on the plantar surface of the metatarsus (28). The metatarsus is painful and swollen and digital pressure exudes a thick purulent exudate.
a What is your diagnosis?
b What rabbit breeds are commonly affected?
c How would you treat the condition?

27 a The chinchilla is a hystricomorph rodent originally found in the high Andes Mountains of South America. This species was hunted to near extinction in the early 1900s and 11 animals were brought to California in 1923. Most chinchillas in the USA are related to these 11 animals. Until recently, chinchillas were bred for their pelt or medical investigation (hearing research and studies of Chagas disease). It is now bred for the pet trade.
b Although chinchillas survive in a variety of environments, they do best in cool, dry conditions. Cold temperatures are preferred in the pelt industry as it stimulates a thick coat. Heat stroke is a significant problem, especially in humid environments. Wire cages are commonly used but solid flooring in a portion of the cage reduces foot problems. Since chinchillas enjoy climbing and jumping, recommend a multilevel cage for pet animals or research animals that are kept for long periods of time. Commercially raised chinchillas are typically caged in polygamous groupings. If a single large cage is used, provide escape boxes to protect other chinchillas from the aggressive behavior of the dominant females. Chinchillas require a dust bath to keep their fur coat groomed. If they are deprived of this, the coat will appear oily and matted and the chinchilla may become depressed and anorectic. Place a mixture of silver sand and Fuller's earth (9:1) in a small container in the cage for a short period of time each day. If left in the cage, the bath will be soiled with feces and food.
c In females, the cone-shaped clitoris is easily confused for a penis. The female has four mammary glands, one at each inguinal region and one on each lateral rib. There are two uterine horns opening into one cervix. A membrane closes the vaginal vault except during estrus and parturition. In the male, there is no true scrotum. The testicles are located in the inguinal canal area and are palpated just beneath the skin. This necessitates closure of the inguinal canal when castration is performed. Determine the sex of young chinchillas by measuring the length of the perineal distance, which is longer in males than females.

28 a Ulcerative pododermatitis or 'sore hocks' is the most likely diagnosis.
b Breeds with thin fur on the foot pads. These include the Rex, New Zealand, Satin, Californian, Flemish Giant and Checkered Giant breeds.
c There are a variety of factors that result in pododermatitis in the rabbit. A purulent exudate indicates the lesions are secondarily infected with bacteria. Rabbit pus is thick due to the lack of proteolytic enzymes in heterophils. To treat this condition, improve husbandry, improve diet, treat the bacterial component and make the rabbit more comfortable. Provide the rabbit with a clean, non-wire cage floor. Take radiographs to determine the extent of the lesion and to help to determine how deep to debride. Debride necrotic tissue while the rabbit is anesthetized to minimize pain. Obtain microbiological cultures from deep lesions. Debride lesions as often as necessary to allow healthy tissue to form. Apply bandages and change them daily if the lesions are extensive. Wet to dry bandaging removes necrotic debris during early treatment of the lesions. Soak the wounds in an antiseptic solution one to three times a day. Apply a topical antibacterial agent after soaking. Administer systemic antibiotics based on culture results in cases where the lesions are extensive.

29 A young rat develops tail lesions. There are bleeding ulcers along the length of the tail and the tissue between the ulcers is necrotic and peeling. Some areas of the tail are edematous while others are constricted. Distal to the constrictions, the tissue is dark and non-bleeding.

a What is the term used to describe this condition?
b What is the etiology?
c How would you treat the condition?
d What should the owner do to prevent this outcome in the future?

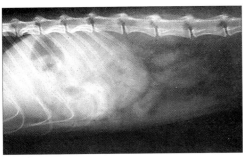
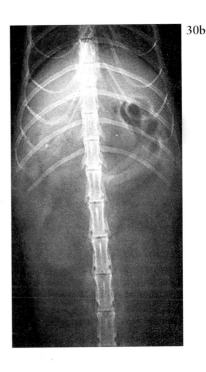

30 A one-year-old spayed female ferret housed indoors is experiencing decreased appetite with intermittent diarrhea and melena of three weeks duration. The ferret is depressed, thin and weak. The mucous membranes are pale and the hematocrit is 27%. Lateral (30a) and ventrodorsal (30b) radiographic views are shown.

a What is your diagnosis?
b What differential diagnoses would you consider in a one-year-old ferret with these signs?
c What other diagnostic tests might be helpful?

29, 30: Answers

29 a Ringtail. Rats with ringtail can have one or more annular constrictions on the tail. This results in edema, inflammation and necrosis distal to the constrictions. Ringtail is a disease of young, weanling or neonate rats.
b Chronically poor husbandry is the proposed etiology. Decreased humidity (<20%) is one part of the problem. Housing animals in wire mesh bottom cages with hygroscopic bedding in rooms with excess ventilation exacerbates this problem. Ringtail may be caused by an abnormal response to environmental conditions or poor temperature regulation of the tail.
c Treat the tail based on the severity of the disease. Debride necrotic material and change the bandages daily if necessary. Daily hydrotherapy appears to be soothing. Apply antibacterial topical medication as needed.
d Correct any improper husbandry conditions: house the rats with proper humidity, bedding should not be excessively hygroscopic and rats should be examined daily to look for early signs of this disease.

30 a Gastric foreign body is the most likely diagnosis. The radiographs show a gastric radiopaque density. A rubber band was surgically removed from the stomach and the ferret recovered uneventfully. There was a small gastric ulceration found near the pylorus that was probably responsible for the recurring melena.
b GI foreign body is the most common diagnosis in a young ferret that is inappetent or anorexic. Vomiting is not usually associated with a chronic gastric foreign body but may be present in cases where complete GI obstruction occurs. Chronic gastric foreign bodies cause gradual loss of body condition due to decreased appetite over time. Chronic bleeding ulcers result in a blood loss anemia. The foreign object or a focal area of abdominal pain can often, but not always, be palpated in affected ferrets. Grasp the ferret firmly by the loose skin along the back of the neck (scruffing) and allow the body to hang unsupported to palpate the cranial abdomen thoroughly. Rubber foreign bodies are most common in ferrets under one year of age. Gastric trichobezoars are seen most often in ferrets over two years of age. An enlarged gastric lymph node can be mistaken for a gastric foreign body on palpation.
  Other possible differential diagnoses in a ferret this age showing these clinical signs are, in descending order of frequency: lymphoma, *Helicobacter mustelae* gastritis (which may result in gastric ulcers), eosinophilic gastroenteritis, GI tract polyps, proliferative bowel disease, Aleutian disease virus infection, renal failure secondary to polycystic renal disease or toxins.
c Because GI foreign body and lymphoma are the two most likely diagnoses in this case, the most important tests to perform are whole body radiographs (two views) and a CBC. Take radiographs after a four to six hour fast to empty the GI tract of its contents. Perform abdominal ultrasound, barium radiography or endoscopy if radiography or abdominal palpation is inconclusive. Evaluate metabolic status with serum biochemistries and urinalysis.

31 The rabbit has unique GI anatomy and physiology.
a How does the 'digestive strategy' of the rabbit differ from other herbivores?
b What is unique about the stomach of the rabbit?
c Why are suckling rabbits predisposed to bacterial diseases of the GI tract?

2a

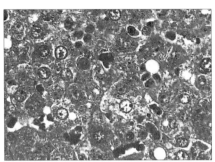

32b

32 An institution has a collection of marmosets and tamarins including this golden lion tamarin (**32a**), which is an endangered species. Several animals are acutely weak and anorectic and some are exhibiting jaundice. They have a mild lymphocytosis along with an increase in both AST and serum bilirubin. Several have died. Gross necropsy findings include jaundice, pleural and pericardial effusions, subcutaneous and intramuscular hemorrhages, splenomegaly and enlarged, yellow-tan livers. Histopathology of the liver reveals diffuse hepatocellular necrosis, acidophilic bodies and a mild inflammatory infiltration (**32b**) (photograph courtesy of Richard Montali).
a What is your diagnosis?
b What is the underlying etiology and is it zoonotic?
c What would you recommend to prevent future exposure?

31 a Rabbits are herbivores and hindgut fermentors like horses (colon fermentor) and large ruminants. However, because of their small body size, rabbits are unable to store large volumes of coarse fiber for long periods of time which would allow for bacterial and protozoal digestion. Rabbits have a system that eliminates fiber from the gut as rapidly as possible and employs its digestive process on the non-fiber portion of forage, which is directed into the cecum for fermentation. This system is driven by the presence of fiber in the diet.
b The rabbit has the largest stomach and cecum of any monogastric mammal. The stomach of the rabbit is simple and acts as a storage vessel for the ingested feed. The rabbit has a well-developed cardiac sphincter which is arranged in such a way that the rabbit cannot vomit. Adult rabbits have a remarkably low pH, between 1.5 and 2.2, in their stomach.
c Suckling rabbits have a pH of 5–6.5 in their stomach. The pH drops at the time of weaning. The higher pH allows the normal microbial population of the gut to develop; however, it also predisposes the young rabbit to develop pathogenic bacteria infections.

32 a These monkeys are dying from callitrichid hepatitis which is a sporadic epizootic in marmosets and tamarins.
b Lymphocytic choriomeningitis virus (LCMV) which is an arenavirus. LCMV is a common rodent pathogen. Wild mice of many species are the principal reservoir hosts but laboratory mice and hamsters serve as important transmitters to other animals and people. Asymptomatic rodents shed LCMV. The virus is shed in urine, saliva and milk. Vertical transmission is common in infected mice. There are two forms of LCMV infection in mice. One form is acquired perinatally and results in a persistent, asymptomatic infection with life-long viremia and viral shedding. These animals are asymptomatic until 7–10 months of age when an immune-complex glomerulonephritis develops. The mice then appear emaciated with a roughened hair coat and a hunched posture and die. In the other form, mice acquire LCMV after the first week of life. This is the non-tolerant or acute form. There is viremia but no viral shedding. The mice either die acutely or recover by eliminating the virus via circulating antibodies. Humans may contract LCMV from contact with carrier rodents. Clinical manifestations in people include fever, headache, myalgia, nausea, vomiting, sore throat and photophobia.
c Discontinue the practice of feeding neonatal mice to non-human primates and eradicate wild mice from cages housing callitrichids. Screen any rodents caught within a primate facility via IFA or ELISA serology for the presence of LCMV.

3a

33b

33 A four-year-old castrated male ferret that is lethargic, has a decreased appetite and a swelling of the right side of the face (33a). The swelling is soft and non-painful; it bulges into the oral cavity with a 1 cm wide, round translucent area in the center resembling a fluid pocket (33b).There are no other oral abnormalities. The ferret is slightly tachypneic with harsh airway sounds and a left-sided 4/6 holosystolic murmur.
a What is the etiology of the swelling?
b How would you diagnose and treat the condition?
c What is a possible cause of lethargy and inappetence in this ferret?

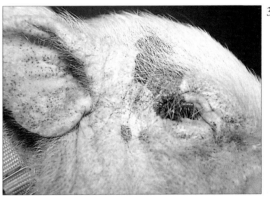
34

34 A young potbellied pig is observed rubbing its neck and ears. The physical examination shows crusts and erythema of the face, neck, margins of the ears, axilla and limbs (34).
a What is your diagnosis?
b What are the differential diagnoses?
c How would you confirm your diagnosis?
d What treatment would you recommend for this pig?

33 a Salivary mucocele is the most likely etiology. This condition is not common in ferrets, but has been reported in one- to two-year-old animals. Other common causes of facial or mandibular swelling include enlarged lymph nodes and abscesses.
b Perform a fine-needle aspirate of the swelling which will reveal a mucoid to tenacious, clear or blood-tinged fluid. Cytology of the fluid shows amorphous debris and occasional red blood cells. Lance or marsupialize the lateral wall of the mucocele. Recurrence is more likely with lancing than with marsupialization. Surgically excise the affected salivary gland if recurrence is a problem.
c Salivary mucoceles in ferrets are rarely associated with clinical signs other than swelling. A possible cause of this ferret's inappetence and lethargy based on the signs present is cardiac disease. Diagnose and characterize cardiac disease with radiographs, an electrocardiogram, and an ultrasound. The other most common cause of lethargy, particularly if it is intermittent, in a ferret of this age is hypoglycemia caused by an insulinoma (beta cell tumor of the pancreas). One or more fasting blood glucose samples should be assessed. Other causes in the aged ferret include adrenal disease, metabolic disease and neoplasia (particularly lymphoma). Simultaneous occurrence of multiple medical conditions is very common in ferrets, particularly in animals over three years of age.

34 a Sarcoptic mange (*Sarcoptes scabiei*) is the most likely diagnosis. Infested pigs have scales, crusts and erythema which begin on the face, pinna, axilla and distal extremities but develop to encompass much of the skin in chronic cases. This disease has zoonotic potential and owners may develop transient erythematous lesions on their skin, especially on the inside of their arms and on their chests.
b Occasionally, in the USA, the 'stick-tight' flea of poultry (*Echidnophaga gallinacea*) is seen on potbellied pigs. The demodex mite of pigs (*Demodex phylloides*) is typically not pruritic. Although lice may have roughly the same distribution on the pig as sarcoptes, the pig louse (*Haematopinus suis*) which is 3–6 mm in length and gray-brown in color is easily identified.
Other causes of erythema and crusting include any source of localized irritation. Sunburn commonly occurs on the unpigmented skin of pigs exposed to long periods of sunlight. If severe, these lesions are followed by pain, flaking, cracking and hemorrhage. Erythema is also caused by other irritants, such as extreme cold (frostbite) and chemicals. The thick keratin crusts without pruritus, referred to as 'parakeratosis', may be related to several dietary deficiencies or oversupplementation problems. Zinc deficiency, hypovitaminosis A, thallium poisoning, or a diet too high in calcium results in these scales and keratin crusts. 'Dorsal pyoderma' is an exudative, painful bacterial dermatitis most often occurring on the backs of pigs. These lesions are analogous to 'hot spots' (moist dermatitis) in the dog.
c Perform deep skin scrapings from affected areas to look for mites. The scraping should be deep enough to cause slight hemorrhage. Negative skin scrapings do not rule out mange. Response to therapy is often the best diagnostic test. Use skin biopsies taken from several locations to confirm the diagnosis. In this case, the diagnosis was sarcoptic mange based on the results of a skin scraping.
d Ivermectin (0.2–0.4 mg/kg SC) repeated once or twice at 10–14 day intervals.

35 A three-year-old female guinea pig has tissue protruding from the vulva (35). The pig farrowed last night and there were three dead pups present in the cage in the morning.

a What is your diagnosis concerning the protruding tissue?
b What treatment would you recommend?
c How would you perform the procedure?

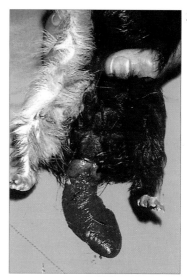

36 The tibial fracture in a chinchilla was stabilized with an intramedullary (IM) pin and then a type I external skeletal fixation device was added (36).

a Why was the external fixator added to the IM pin?
b What is the minimum number of pins that should be used with an external fixator device when an IM pin is also present?
c What is the minimum number of IM pins required if the external fixator device were the sole means of stabilization?
d Why are the pins inserted at an angle with respect to each other rather than parallel to each other?
e When are the implants removed?

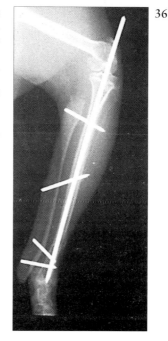

35, 36: Answers

35 a Uterine prolapse.

b Ovariohysterectomy is the treatment of choice in this case. Stabilize the patient metabolically before anesthesia and surgery. If the animal is debilitated, reduce the tissue and place a purse-string suture as a palliative procedure prior to definitive surgical management.

c The technique of ovariohysterectomy in guinea pigs is similar to that in rabbits though the uterus is not coiled. Place the patient in dorsal recumbency and prepare for aseptic surgery. Make an incision about 1–2 cm long just cranial to the pubis. Identify the uterine horns dorsal to the apex of the bladder. Grasp one horn and exteriorize it through the incision. Trace the horn craniad to the ovary. The oviduct circles cranially all the way around the ovary. The ovaries, located at the caudal pole of the kidneys, are within a large fat pad. As in rabbits, the suspensory ligaments are long and it is easy to exteriorize the ovaries. There is a single artery and vein which run medial to the ovaries and extend along the uterus following the uterine horns to the uterine body. Place a clamp between the ovary and the uterine horn to allow traction to be applied to the ovary. The ovarian ligament need not be broken down. There are many vessels which supply the ovary within the fat of the mesovarium. Create an opening by blunt dissection through the fat of the mesovarium and pass a ligature around the portion of the mesovarium containing the vessels supplying the ovary. Tightening the suture cuts through the fat and ligates the blood vessels. Alternatively, place a vascular clip to control hemorrhage. Repeat this procedure on the contralateral side. Break down the fat-filled broad ligament of the uterus by gentle dissection. Ligate or place hemoclips on any large vessels or any hemorrhaging vessels within the broken ligament. Ligate the uterus just cranial to the cervix to prevent urine leakage. Care is taken to avoid damaging the urinary bladder. Closure is routine. Use 4–0 synthetic absorbable suture material.

36 a A single IM pin does not stabilize rotational forces. A transverse fracture, as depicted here, needs rotational stability for proper fracture healing. The external fixation device provides rotational stability.

b Only one pin is required on each side of the fracture when using an external fixation device to add rotational stability. With one fixation pin proximal to the fracture and one distal, the fracture cannot rotate. Add additional pins for more stability.

c When using an external fixator device as the sole means of fixation, there must be at least two fixation pins (although three are preferred) proximal and distal to the fracture site.

d To prevent the device from being easily pulled out of the bone. If all pins are placed parallel to each other, the fixator easily pulls off.

e As with any fracture, remove orthopedic implants only after there is radiographic evidence that the fracture is healed. There is no set time when this occurs.

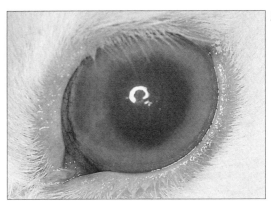

37 A six-month-old New Zealand White rabbit appears to be blind in both eyes. The figure (37) shows a white ring visible at the limbus. The rabbit is one of a litter of four purchased from a laboratory colony to be a pet. The rabbit had apparently normal vision until recently.
a What is your diagnosis?
b Is this likely to happen to the other rabbits in the litter?
c Are there any other problems associated with this condition?

38 A two-year-old potbellied pig has caught its hind limb in the doorway of its cage and tears off the rear claw. To repair this surgically, the pig must be anesthetized.
a Which injectable anesthetics can be used for tranquilization or pre-medication in potbellied pigs?
b If an inhalant anesthetic is chosen, which is most commonly used in the potbellied pig and why?
c Which procedure would you use to intubate the pig?
d What is the greatest problem of using analgesics in potbellied pigs?
e Which analgesics would you use in the potbellied pig?

37, 38: Answers

37 a Early corneal edema resulting from raised intraocular pressure due to glaucoma.
b This is an inherited condition in this breed of rabbits. The bu gene causing buphthalmos is quite widespread in the laboratory population of the New Zealand White rabbit. The gene is recessively inherited. Litters produced by parents with normal vision but carrying the bu gene will have an average of one in four of the offspring develop the condition. Animals homozygous for bu are not always affected because the gene is only partially penetrant. Since we do not know the genetics of the parents, it is difficult to predict if the other three rabbits will develop this problem.
c Glaucoma in rabbits does not appear to cause the same painful ocular sequelae that occurs in the dog or cat. Usually, by the time the owner has noticed changes in the eye the animal is already blind. The affected eyes do not become severely buphthalmic, therefore secondary trauma due to an exposed bulging eye is uncommon. Surgery to relieve intraocular pressure can be performed early in the disease in an attempt to prevent blindness, but late in the disease it is of questionable value.

The bu gene also carries with it some semi-lethal effects, such as low litter size (note that only four animals were born in this litter), high infant mortality and occasional sudden death. Recommend that owners not use these animals for breeding.

38 a Use acepromazine maleate (0.05–0.25 mg/kg IM) for reliable sedation. If available, administer azaperone (Stress-Nil, Pitman Moore, Mundelein, IL; Stresnil, Janssen Animal Health) (2 mg/kg IM) for tranquilization for minor procedures. Use ketamine alone (20 mg/kg IM) or ketamine combinations for tranquilization. Administer combinations, such as ketamine (20 mg/kg IM) and acepromazine (1.1 mg/kg IM) or ketamine (20 mg/kg IM) and xylazine (2 mg/kg IM).
b Isoflurane. Both induction and recovery times are less with isoflurane than any other commonly used inhalant anesthetic. Give injectable premedication to intractable pigs. To control excessive salivation, administer atropine (0.05 mg/kg SC). Reportedly, halothane is associated with problems of increased respiratory secretions, catecholamine-induced arrhythmias and hyperthermia. Hyperthermia is also reported as a complication of the use of isoflurane in potbellied pigs but to a lesser degree than with halothane.
c Intubation of the potbellied pig is challenging. The epiglottis lies behind an elongated soft palate and a tracheal tube is easily misplaced into the laryngeal diverticulum. Intubate the pig while it is in lateral recumbency. Use a spring-type oral speculum to hold the mouth open, fully extend the tongue and use a 150–195 mm laryngoscope blade to lift the palate. Place a stylet in the endotracheal tube to aid in its passage. Introduce the tube through the arytenoid cartilages and twist it as it is advanced through the vocal folds into the airway.
d Most analgesic agents have a very short half-life in the potbellied pig.
e The two longest-acting analgesic agents in swine are butorphanol (0.1–0.3 mg/kg q 4–6 hours) and buprenorphine (0.05–0.10 mg/kg IM q 8–12 hours). Other narcotic agents have half-lives of less than one hour and must be given by constant infusion to be effective.

39 Many diagnoses would not be possible in small rodent pets without appropriate laboratory samples. Blood collection from the orbital sinus is not acceptable in the pet rodent because of technique complications, such as nasal or ocular blood draining or blindness.
a Where can you collect blood in the mouse, rat, gerbil and hamster?
b How would you collect blood from these sites?

40 It is common and normal for rabbits to have a thick, white urine due to calcium carbonate precipitate (**40**). It is also common to observe calcium concentrations between 3–4 mmol/l (12–16 mg/dl) in the serum of normal rabbits.
a What is unique about calcium metabolism in the rabbit?
b What factors could increase calcium carbonate precipitation in the urine?

39

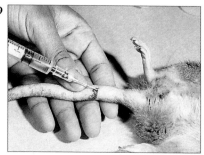

39 a Mice, rats and gerbils all have large lateral tail veins (39); hamsters have a ventral caudal tail artery which is more accessible in the female. Jugular veins and femoral arteries, while not easily visualized, are equally as large as the tail veins. Cephalic and lateral saphenous veins are more easily visualized, but are smaller in diameter. Do not use a toe nail clip to collect blood as it is painful, stressful and the blood can mix with tissue fluids leading to inaccurate results.

b It may be necessary to anesthetize some patients for successful phlebotomy. Excessive restraint in debilitated animals can have fatal results. Slightly warm the patient before venipuncture to dilate the vessels. Calculate the blood volume to be collected. In a hydrated, healthy animal, 6–10% of the body weight in grams is the total blood volume. A maximum of 10% of the total blood volume may be taken in whole blood. Use an insulin syringe to reduce blood waste in the needle hub and prevent excessive negative pressure on the vessel. Break the suction on the syringe and coat the syringe with heparin before use. Allow blood to flow into the syringe without pulling on the plunger until the vein has time to fill again. Consider hemorrhage as part of the total collectible volume. Apply pressure to the vein until hemostasis is acheived.

40 a Rabbits are unique in their methods of absorption and excretion of calcium. Many species of animals absorb and control calcium in relation to their metabolic needs, regulated by the parathyroid hormone interacting with metabolites of vitamin D and calcitonin. Rabbits appear to absorb calcium from the gut in direct proportion to the concentration in the diet. This absorption appears to be independent of metabolic need. It is still unclear as to the role that vitamin D plays in calcium absorption in the rabbit. The primary route of calcium excretion is renal. When the blood calcium level exceeds the kidney threshold, the excess blood calcium is excreted in the alkaline urine. This is different than other species where the bile is the primary route of removal. Plasma calcium concentrations can change dramatically depending on the amount of calcium in the diet. 'High' plasma calcium concentrations from 3–4 mmol/l (12–16 mg/dl) are commonly found in rabbits on a calcium-rich diet, such as those eating alfalfa-based pellets and alfalfa hay.

b The two most common factors that will result in increased calcium carbonate precipitate in the urine are increased dietary calcium and an increase in the urine pH. The form of the calcium in the diet also influences its digestibility. For instance calcium oxalate is about 49% digestible, dicalcium phosphate is about 53% digestible and calcium carbonate is about 81% digestible.

The rise of urine pH may be associated with urinary tract infection. In addition, inadequate water intake can obviously lead to a more concentrated urine. The role vitamin D plays in calcium metabolism in the rabbit is controversial. Some studies indicate that high dietary concentrations of vitamin D along with high dietary calcium concentrations may predispose the rabbit to calcification of the aorta and kidney, but this is not always reproducible. For this pathology to occur, it may be necessary to maintain high concentrations of vitamin D and calcium for prolonged periods or there may be other unknown metabolic or dietary factors involved.

41 An eight-year-old female owl monkey fell from a branch about 12 m above the ground. Two days following the fall, it is still lame on the right hind limb. A radiograph of the pelvis is taken (**41a**).
a What fractures do you see?
b What would you recommend for the surgical management of these fractures and why?

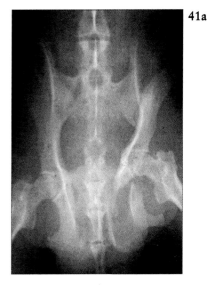

41a

42 In the USA, pet ferrets are routinely vaccinated for rabies and canine distemper virus. What vaccine schedules would you recommend for the ferret?

43 A pet rabbit has signs of an upper respiratory tract infection and is exhibiting a head tilt with occasional circling. The photograph shows ocular lesions observed in this patient (**43**).
a What ocular lesions are present?
b What is the likely cause?
c How would you treat this condition?

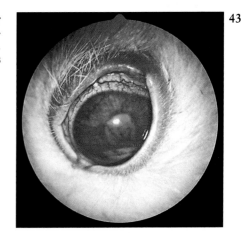

43

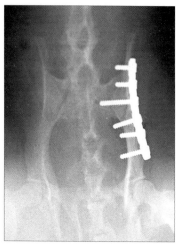

41b

41 a There is a long oblique fracture of the ilium, a fracture of the ischium and a pubic fracture on the right side which creates a free-floating acetabular segment.

b Repair these fractures with a 2.7 mm reconstruction plate applied to the ilium (**41b**). This pulls the ischial and pubic fractures into alignment. Due to both of the large muscle masses that surround the pelvis and the ample pelvic blood supply, the pubic and ischial fractures do not require further stabilization. Additionally, unlike the ilium, they are not along the weight-bearing axis. The ilium plate stabilizes the one fracture that is along the weight-bearing axis, allows rigid fixation and gives anatomic alignment for early return to function. This is vital for non-human primates as postoperative activity restriction is difficult.

42 a Recommend vaccination against canine distemper virus because the disease in ferrets is nearly always fatal. In the USA, the canine distemper virus vaccine licensed for use in ferrets is Fervac-D (United Vaccines, Madison, WI). This is a modified live vaccine given subcutaneously. Administer the first vaccination at six to eight weeks of age and repeat boosters at two to three week intervals until the ferret is 16 weeks of age and then booster annually. Ferrets over 16 weeks of age that have never been vaccinated receive two boosters two to three weeks apart. Do not use a canine distemper virus vaccine that is propagated in a canine cell line because it can result in clinical disease.

Recommend vaccination against rabies virus in the domestic ferret. Although the risk of exposure is small in ferrets housed indoors, local animal control ordinances may require vaccination. In some areas, an unvaccinated ferret that bites a human will be euthanatized and the head submitted for rabies virus examination. Not every municipality will recommend vaccination. In the USA, the killed rabies virus vaccine licensed for use in ferrets is IMRAB-3 (Rhone-Merieux Inc., Athens, GA). Ferrets are susceptible to rabies virus but are not natural reservoir hosts of the virus. The incubation period for rabies virus in the domestic ferret is still under investigation. Vaccinate ferrets subcutaneously at three months of age and give boosters annually. Ferrets do not need to be vaccinated against feline panleukopenia, feline leukemia and feline infectious peritonitis viruses because they are not susceptible to these diseases.

43 a The ocular lesions visible are hypopyon, aqueous flare, anterior synechiae, iridal neovascularization and episcleral congestion.
b These are all signs of anterior uveitis. Microbiologic culture of an aqueous tap was negative. *Pasteurella multocida* was cultured from a ventral bullar osteotomy. The ocular lesions probably arose from hematogenous spread of the organism.
c Treat this ocular condition with a combination of mydriasis with atropine, rigorous topical anti-inflammatory medication and systemic and topical antibiotics.

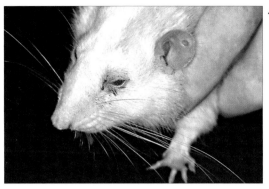

44 An adult male rat in a pet shop has what appears to be blood coming from his eyes (44). Two other adult males in the same cage are sneezing and sitting hunched in the corner with rough appearing hair coats. There is a total of six large rats in the 75-litre glass tank with fresh pine shavings for bedding.

a What is the red staining around the rat's eyes?
b What is the likely etiology for the rats' condition?
c What predisposing factors are associated with rat respiratory diseases?

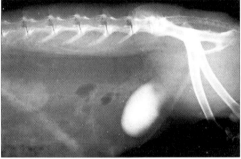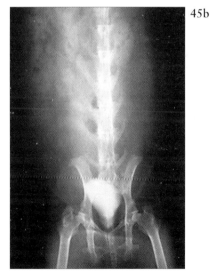

45 An owner reports dysuria in a rabbit for the last three days. Radiographs are taken and radiopaque material is seen in the bladder (45a, b). Urinary tract calculi of varying sizes and abnormally large amounts of calcium carbonate (commonly referred to as 'sludge') are common in the rabbit.

a What are the signs of urinary tract calculi or sludge in the rabbit?
b How is this diagnosed?
c How is urolithiasis treated in the rabbit?

44 a Porphyria. It occurs when the rat's Harderian gland becomes inflamed and secretes porphyrin-pigmented tears.

b A common viral agent, SDAV, is frequently responsible for this highly contagious condition, which is called sialodacryoadenitis. SDAV is an RNA virus that replicates in the epithelial cytoplasm of the respiratory tract and travels along the ducts into the glands of the head. It has a special predilection for the Harderian gland. Younger animals are more severely affected and the virus usually resolves in a week.

Other causes of porphyria include infections caused by Sendai virus, P3 virus and corona virus. Organisms such as *Pasteurella pneumotropica* and *Mycoplasma pulmonis* can act as common bacterial secondary invaders. It is advisable to treat most cases with antibiotics following appropriate cultures and sensitivities. In some cases the porphyria persists due to permanent damage to the Harderian gland.

c Bedding, such as cedar or pine shavings or cypress mulch, can predispose rats to respiratory disease. They contain the volatile oil thujone. Thujone is a respiratory irritant and may be tumorogenic. Thujone can cause convulsions and cortical brain lesions if there is prolonged exposure. In addition, volatile oils are strongly scented and may mask feces and urine odors in the cage. In an aquarium or other enclosed environment, ammonia and other toxic gases, which are heavier than oxygen, sink to the floor where the animals are living. This disastrous situation may cause respiratory conditions to deteriorate rapidly. Hardwood shavings or pellets are a much better choice for cage litter, with no sanitary compromise. Other predisposing factors include overcrowding, high environmental temperatures and inadequate nutrition.

45 a They include calciuria with one or more of the following signs: anorexia, dysuria, stranguria, reluctance to move, a hunched posture and perineal staining with calcium carbonate precipitate. If calculi are in a ureter or kidney, these structures may need to be removed and therefore a preoperative radiographic contrast study is recommended to check the function of the opposing kidney.

b With abdominal radiographs, ultrasound and urinalysis, if indicated.

c Treat urolithiasis in rabbits by flushing the urinary tract or perform a cystotomy. Perform hematology, serum biochemistries and a urinalysis. Take a urine microbiologic culture. Since low-grade cystitis may be present, administer antibiotics.

If bladder sludge is present, flush the bladder repeatedly with the rabbit anesthetized. The use of a mildly acidic solution may aid in this process. Gently flush until the urine is clear. Use analgesics postcatheterization as urethral spasm is common. Diurese these patients with parenteral fluids. Hospitalize rabbits until they are urinating normally. If bladder sludge cannot be removed by flushing, a cystotomy is then necessary.

Perform a routine cystotomy to remove urinary tract calculi or sludge that cannot be removed with flushing. Culture the bladder mucosa and analyze the calculi for mineral composition. Closure is routine. Postsurgically, administer analgesics.

Perform follow-up radiographs and urinalysis routinely to check for recurrence. If needed, increase water consumption by the addition of fruit-flavored drink additives (e.g. fruit juice, oral electrolyte solutions) to the water. Reduce dietary calcium to decrease the amount of calcium excreted in the urine. Prevention of recurrence is hampered because the pathology of this disease is not yet fully understood, therefore advise owners that this disease, even with appropriate management changes, may recur.

46 A female potbellied pig needs to have an ovariohysterectomy.
a What is the preferred age to perform this procedure?
b What are the landmarks for the skin incision?
c How is the conformation of the uterus different from that of dogs and cats?

47a

47 A squirrel monkey has a large swelling on its head (47a) (photograph courtesy of J.L. Wagner). The monkey is sensitive to palpation in many joints and is reluctant to move. Swellings are apparent over the epiphyses of the long bones and are seen radiographically. The monkey eats a commercial diet, purchased over a year ago, formulated for New World monkeys. The feed is stored in a plastic garbage can. The diet is supplemented with apple sauce, fruit cocktail and treats such as marsh-mallows.
a What is the cause of the swellings and reluctance to move?
b How would you treat the condition?
c What would prevent recurrence?

46 a Eight weeks of age for a number of reasons. In a prepubescent gilt, the uterus is small and coiled with small vasculature. In the adult sow, it is larger and prominent but with large blood vessels. Although it is easier to identify the uterus in an adult, the larger vascular supply makes for a more difficult surgery. The underdeveloped vascular supply in gilts minimizes the risk of hemorrhage. Also, breeders often prefer to place young pigs that have already been spayed. Ovariohysterectomy obviously prevents estrus. Once pigs reach puberty, they are polyestrus and continue to cycle until they get pregnant. Owners find this annoying, especially in house pets.
b Make the incision on the ventral midline between the most caudal two pairs of teats. Following the body wall incision, identify the uterus. In a prepubescent gilt, the uterus is tightly coiled and lays between the colon and urinary bladder. Use a spay hook to help locate the small uterus of a gilt.
c The ovaries are located close to the uterine body so the horns are tightly coiled and not laid out in a 'Y' configuration as in dogs and cats. This makes identification and ligation of the blood supply more tedious. Also, in pigs the ovaries are less firmly attached and the uterus is more friable.

47b

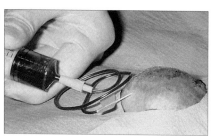

47 a Vitamin C deficiency or scurvy. This leads to cephalohematoma, joint pain and skeletal changes, such as swelling of the epiphyses of long bones. Additional signs include periosteal hemorrhage and gingivitis associated with damage to connective tissue surrounding the teeth. In severe cases, hemorrhages occur throughout the body. Vitamin C deficiency is common in New World monkeys on poor or marginal diets. New World monkeys are one of the few groups of animals that have a need for vitamin C supplementation. Since this monkey was consuming a diet formulated for New World monkeys, it should have had adequate vitamin C concentrations. However, vitamin C is labile, being sensitive to the damaging effects of heat and light. The vitamin C was no longer active in the packaged diet. In addition, there were no other dietary sources of vitamin C in the diet.
b Correct the underlying dietary problems. Administer vitamin C (ascorbic acid) (25 mg/kg/day IM, SC or PO). Most non-human primates readily accept chewable vitamin C tablets. Feed citrus fruits and other foods high in vitamin C. Drain cephalohematomas using aseptic technique with a 19 or 21 gauge butterfly catheter (47b) (photograph courtesy of Dr. J.L. Wagner). Either place drains in the subcutaneous space and cover with bandages or drain the hematoma daily for several days. If bandaged, change dressings daily. Administer an analgesic, such as acetaminophen (5–10 mg/kg PO q 8 hours) to help ease joint pain and discomfort. Soften foods to decrease oral trauma if the teeth and gums are involved.
c By maintaining adequate amounts of vitamin C in the diet. The potency of vitamin C decreases rapidly if exposed to excessive heat or moisture. Most commercial New World monkey diets lose their vitamin C potency within 90 days after milling. Supplement the diet with fresh fruits and vegetables high in vitamin C. Also, add vitamin C tablets to the diet to maintain a concentration of vitamin C of approximately 2 mg/kg/day. Use 25 mg/kg/day when deficiencies are present.

48 A 16-week-old male rabbit has a right-sided head tilt, vertical nystagmus (fast phase up) and mild epistaxis (48). The rabbit was left in the backyard to exercise a few hours before the examination. It panicked and ran forcefully into the fence when a dog entered the yard. Postural reactions are difficult to assess due to the rabbit's tendency to fall to the right, but pain and proprioception appear to be normal. The right eye lacks

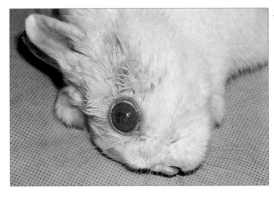

menace and palpebral responses, but has a normal PLR. Although the rabbit is breathing rapidly, on auscultation of the thorax all parameters are within normal limits. Otoscopic, fundic and oral examinations are within normal limits. The rest of the physical examination is unremarkable.
a Localize the lesion.
b What emergency treatment plan would you recommend?
c What one additional diagnostic test would be most useful in determining a prognosis?

49 A two-year-old male gerbil has crusts around the eyes, is depressed and anorectic (49). The perinasal area is erythematous with patches of alopecia. Both front paws are denuded medially and the hair coat over the entire body is matted.
a What is responsible for these signs?
b What factors contribute to nasal dermatitis in the gerbil?
c How would you manage this condition?

48, 49: Answers

48 a The right-sided head tilt and tendency to fall to the right indicates a lesion to the right vestibular system, either central or peripheral. The vertical nystagmus with fast phase up indicates a central lesion most likely in the central vestibular region. Localization of cranial disease is hampered in this species because palpebral and menace responses are often blunted when rabbits are stressed or in shock.
b Treat for shock and head trauma. Establish an intravenous or intraosseous catheter. Administer sodium prednisolone succinate (5–10 mg/kg IV or IO) or dexamethasone sodium phosphate (4 mg/kg IV or IO). Administer lactated Ringer's solution (approximately 20 ml/kg over 1 hour) to control hypovolemia, but do not overhydrate because this will contribute to cerebral edema. Provide oxygen by cage, mask or nasal catheter to reduce cerebral edema. Ensure that the rabbit is ventilating adequately. Furosemide (2 mg/kg IV or IO) may aid in decreasing production of cerebrospinal fluid. Lubricate the right eye to protect it from desiccation.
c A skull radiograph to rule out fractures. The epistaxis and CNS signs could be caused by soft-tissue trauma, which would have a good prognosis for recovery over 10–14 days. If skull fractures are present, the prognosis is guarded for non-displaced fractures. In these cases the potential for intracranial hemorrhage is greater. It may be necessary to perform cranial surgery to relieve the pressure. Displaced skull fractures resulting in malocclusion of the teeth must be considered in the long-term prognosis of the rabbit.

49 a The Harderian gland. This is a periocular lacrimal gland found in many small rodents. It drains into the nose and when inflamed its secretions increase. The lacrimal secretions contain a porphyrin pigment (red-orange in color) that is irritating to the skin. Alopecia on the face and front legs is caused by the gerbil trying to clean the excessive secretions from its eyelids and nose. In addition, overgrown, sharp or broken toenails can cause ocular damage including corneal lacerations. Secondary loss of ocular or periocular integrity allows bacterial invasion resulting in conjunctivitis. The gerbil becomes weakened and depressed when it can no longer see or smell its food.
b (1) Bedding that is too rough or too shallow. Gerbils live in burrows and spend a great deal of time digging. Perinasal tissue integrity is lost when thin layers of bedding or substrates, such as sand or rough shavings, are used. *Staphylococcus* spp. and *Streptococcus* spp. infections are often secondary invaders.(2) The gerbil is a desert dwelling species and poorly tolerate a humidity level above 50%. (3) Bedding containing volatile oils, such as cedar shavings. (4) Primary or secondary mycoplasma infections, as well as some viral infections, contribute to disease of the respiratory tract, which may lead to increased nasal and Harderian gland secretions. (5) Stress. Common causes of stress are dirty cages, improper or inadequate diet or an incompatible cage mate.
c Change the bedding to a soft, non-toxic material, such as hardwood shavings or pellets, cellulose pellets or newspaper products. Provide at least a 5 cm depth of bedding. Gently remove crusts and skin debris and cleanse the face with a mild antiseptic solution. It may be necessary to use anesthesia in severe cases. Trim the nails to prevent further abrasions. Apply a non-steroidal ophthalmic antibiotic ointment to the affected areas at least twice daily. Use the minimum amount of topical medication because gerbils will ingest any excess during grooming. Avoid separating compatible gerbils for an extended period of time, because they may fight when reunited.

50 An adult pet mouse is left in the care of a young girl while the owner is on holiday for two weeks. When the owner returns, she finds the mouse near death and takes it to the local veterinary clinic. The young girl is not available for questioning.
a What possible etiologies should be considered in this case?
b What emergency therapy should you institute?
c What precautions should be taken with hospitalized rodents?

51 A four-year-old pet female guinea pig has progressive hair loss for one year. Repeated skin scrapings are negative for parasites. Areas of thinning and alopecia are bilaterally symmetrical over the flank area and the skin appears grossly normal.
a What are your differential diagnoses?
b What additional diagnostic tests would you advise?

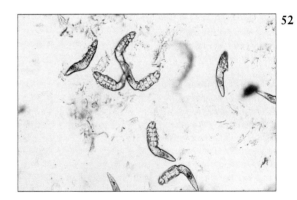

52

52 A skin scraping from a gerbil with a rough hair coat and dermatitis of the face and back is examined.
a What are these organisms (52)?
b Are they involved in the dermatitis?
c How would you treat the organisms?

50 a Lack of food or water, inappropriate food, inappropriate environmental temperatures and injuries due to unskilled handling or escape. It is possible that the mouse was not attended to regularly even if there was food and water in the cage when the owner returned. A history of trauma may be unknown since the caretaker will often put the animal back in the cage without telling the owner of any problem.
b Administer oral electrolyte solutions if the animal can swallow. Warm parenteral fluids to body temperature and administer subcutaneously if the patient is still ambulatory. In cases where more intense therapy is needed, give a bolus of fluid intraosseously through the femur, or intravenously. Take radiographs especially if trauma, GI foreign bodies, heart or lung disease is suspected. Administer oxygen if there is evidence of respiratory distress. Trauma to the limbs, spine or teeth caused by falls from unskilled hands or from enclosure escapes are common.
c Make sure that rodent cages are escape proof. Even a severely debilitated animal will attempt to find an escape route. Secure lids and doors around all edges, and make sure that wire spacing is smaller than the dorsoventral thickness of the animal's skull. To avoid drowning the patient or creating a cold and wet environment, do not use deep dishes of water. Do not use sipper bottles as the sole source of water because some animals are either too debilitated or painful to drink from them or are unfamiliar with their use. Place marbles or large stones in a shallow water bowl to eliminate spillage and the risk of drowning.

51 a They include ectoparasites (*Chirodiscoides caviae* or *Trixacarus caviae*), lice (*Gliricola porcelli* or *Gyropus ovalis*), dermatophyte infection (*Trichophyton mentagrophytes*), bacterial skin infection, malnutrition, hyperadrenocorticism, hypothyroidism and hyperestrogenism.
b Repeat the skin scrapings and examine hairs and skin debris microscopically for dermatophytes as well as parasites. Submit material for a fungal or bacterial culture as indicated. Perform a skin biopsy if scrapings are negative and other physical findings are normal. Careful abdominal palpation can often reveal abnormal enlargements of the ovaries or uterus. Radiograph and/or ultrasound the abdomen if ovarian or uterine disease is suspected. It may be possible to submit serum for hormone assays to differentiate the various endocrine diseases. This case was diagnosed with cystic ovaries and uterine fibrosarcoma (**51**). The cause of the hair loss was most likely hyperestrogenism caused by the cystic ovaries. The haircoat returned to normal within three months after an ovariohysterectomy was performed.

52 a The organisms are *Demodex* mites.
b They cause dermatitis in gerbils and hamsters. These mites burrow into the skin and therefore these organisms may not be found on superficial skin scrapings.
c Use ivermectin (0.2–0.4 mg/kg SC with ranges up to 1.0 mg/kg PO). Repeat this treatment in 10–14 days. Although, mites are treated in other animals with a water-amitraz mixture, hamsters and gerbils can have a fatal reaction to this drug. Demodicosis is rarely a primary disease and underlying disease is usually present.

53 An otherwise healthy four-month-old male rabbit has a closed midshaft fracture of the tibia and fibula (53). The fracture is repaired with a type II external fixator. The rabbit recovers well from surgery and the fracture appears stable. Over the next 24 hours, the leg shows an expected amount of soft-tissue swelling and bruising. The rabbit is not chewing on the fixator, but avoids using the leg. The rabbit is anorectic, but is drinking water and has a normal rectal temperature.

a What is the cause of this rabbit's anorexia?
b What drugs are recommended for use as analgesics in rabbits?

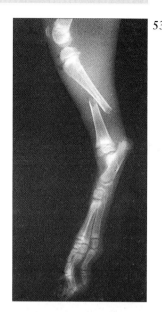

54 A four-year-old ferret is experiencing thinning of the hair on the tail. Upon further inspection a raised, round, ulcerated lesion is noted midway down the length of the tail (the lesion after the hair around it is shaved is shown (54)). The owner reports that the ferret does not appear to be bothered by the lesion. The ferret is on a high-quality commercial cat food diet and is otherwise clinically normal.

a Is the thinning of the hair on the tail related to the lesion?
b What are the differential diagnoses for skin masses in ferrets?
c How would you manage skin masses on ferrets?

53, 54: Answers

53 a Pain. Rabbits are very sensitive to pain, stress or fear and will often become anorectic and depressed when any of these are experienced.
b Butorphanol is a narcotic agonist/antagonist that is considered to be a safe and effective analgesic in rabbits (0.1–0.5 mg/kg IV, IM or SC q 4 hours). Mild sedation is commonly observed at higher doses. Bradycardia, respiratory depression, ptyalism and nausea are potential, but infrequently reported, effects of narcotic agonists/antagonists. Buprenorphine (0.02–0.05 mg/kg SC or IV q 12 hours) is another narcotic agonist/antagonist that has been used successfully in rabbits. Flunixin meglumine is a non-steroidal anti-inflammatory that provides moderate levels of analgesia for visceral and orthopedic pain (1.1 mg/kg SC or IM q 12 hours or 2.0 mg/kg PO q 12 hours). The injectable form of flunixin can be given orally if mixed with a palatable syrup. Acetylsalicylic acid (100 mg/kg PO q 4–8 hours) provides mild analgesia for orthopedic pain. As with any species, it is important to maintain adequate hydration and to monitor for gastric irritation when administering non-steroidal anti-inflammatories.

54 a Not likely. Partial to complete alopecia of the tail can occur in ferrets independent of any other skin lesions or clinical signs. The etiology is unknown. It usually appears at about the time of a normal hair molt. Hair regrowth usually occurs in one to three months. If the hair does not regrow, or if the hair loss progresses above the tail base, adrenal hyperplasia or neoplasia should be considered.
b Neoplasia, enlarged lymph nodes and abscesses. Neoplasms that produce masses in or under the ferret's skin include mast cell tumor or mastocytoma, sebaceous gland adenoma and adenocarcinoma, benign cystic adenomas, fibroma and fibrosarcoma, hemangioma, cutaneous hemangiosarcoma, chordoma, neurofibroma, leiomyoma, histiocytoma, squamous cell carcinoma, basal cell carcinoma, cutaneous lymphoma and melanoma. It is possible to have more than one type of skin neoplasm present on an animal. The diagnosis in this case was mast cell tumor.
c All masses in the ferret should be investigated and surgically excised as soon as possible. Because neoplasia is the the most common cause of skin masses in the ferret, do not recommend a 'wait-and-see' approach. Obtain radiographs and/or an ultrasound to assess for metastatic pulmonary or hepatic disease or underlying tissue involvement. Perform histopathology on the mass to obtain a diagnosis and prognosis.
   Although the majority of skin neoplasms in the ferret are benign, metastasis to distant sites has been reported with mast cell tumors, sebaceous gland adenocarcinomas and squamous cell carcinomas. Chemotherapy and radiation therapy have not been successful in treating metastatic disease.

55 Describe lavage and tube feeding procedures for the anorectic rabbit.

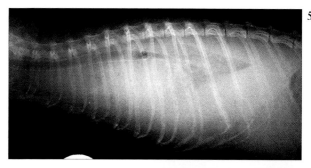

56a

56 A six-year-old male castrated ferret that lives indoors with no other pets is acutely dyspenic and recumbent. It responds well to oxygen therapy. These radiographs are taken (56a, b).
a What are your differential diagnoses for this ferret's respiratory problems?
b Based on the radiographs, what immediate treatment would you consider?
c What other tests would you perform?
d What other treatments would you give?
e What is the prognosis?

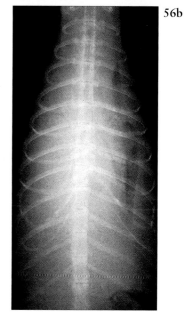

56b

55 Anorectic rabbits can develop hepatic lipidosis relatively quickly. Treat anorexia over the short term with syringe feedings of ground rabbit pellets or alfalfa powder (obtained through health food stores) mixed with puréed fresh vegetables, canned pumpkin or strained human vegetable baby food. Add an oral electrolyte solution to thin the mixture and feed approximately 10–15 ml/kg PO q 8–12 hours.

If prolonged forced alimentation is expected or if the patient is difficult to handle, place a 5–8 Fr nasoesophageal feeding tube. Use good restraint in the conscious rabbit and a small amount of lidocaine gel on the tip of the tube and a drop of lidocaine on the nasal mucosa to decrease any discomfort on placement. Secure the catheter to the skin over the dorsum of the nose and the cranium either with surgical glue or sutures. Use an Elizabethan collar only if necessary (which is rare). Before administering food, radiograph the rabbit's thorax to ensure the tube is properly placed. Rabbits may not cough or otherwise indicate improper tube placement in the trachea when sterile saline is introduced. Administer a high-calorie, non-dairy human or equine liquid food supplement for the short term, or the previously mentioned syringe feeding mixture for the long term. Pass the food through a fine strainer to avoid clogging the nasoesophageal tube. Flush the nasoesophageal tube with saline before and after feedings to maintain a clear line.

Offer the patient fresh hay (grass is preferred) and a variety of fresh greens daily (romaine lettuce, parsley, carrot tops and dandelion greens are often enjoyed). Once eating has resumed with normal stool production, tube feeding can be discontinued.

56 a Heart disease (hypertrophic and dilated cardiomyopathy), neoplasia (particularly lymphosarcoma) and severe influenza infection are the most likely differential diagnoses. Other potential causes of respiratory difficulties (e.g. heartworm, distemper virus) are less likely due to the signalment and husbandry.
b The radiographs reveal thoracic fluid. Immediately treat this ferret by removing the pleural fluid via thoracocentesis. Analyze and culture the fluid. Drain both sides of the thorax by using a 25 gauge butterfly catheter placed low in the chest wall. Aspirate multiple sites on either side of the chest to remove as much fluid as possible.
c Once the ferret is stable, retake the radiographs to view the thoracic contents more clearly. In this case, a large heart is present leading to the diagnosis of heart disease. To further characterize the heart disease, perform an echocardiograph. The echocardiograph determines the type of heart disease present. Dilated cardiomyopathy is the most common form of heart disease in the ferret. Hypertrophic cardiomyopathy is seen infrequently and restrictive cardiomyopathy is very rare. Perform an electrocardiogram to further characterize the heart disease. In this case, the echocardiograph reveals left ventricle dilation, mitral and tricuspid valve regurgitation and a poor shortening fraction. The diagnosis is dilated cardiomyopathy.
d While the ferret is in the hospital, give it oxygen therapy as needed and remove pleural effusion as it accumulates. Begin long-term treatment in the hospital and continue at home. Administer furosemide (2.2 mg/kg PO q 8–12 hours), digoxin (0.01 mg/kg PO q 24 hours) and an ace inhibitor such as enalapril (0.5 mg/kg PO q 48 hours).
e The prognosis depends on the response to treatment. Even ferrets with severe heart failure may do well with treatment and have a good quality of life as long as they continue to respond to the medication.

57 Rabbits have large inguinal rings and canals which allow the testicles to move freely into and out of the abdominal cavity. However, inguinal herniation of abdominal viscera is rare.
a Name the structure pictured which prevents herniation of intestines through the rabbits' inguinal canal (57).
b How are closed and open techniques for rabbit castration performed?

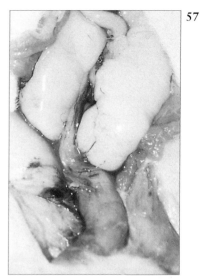

57

58 A one-year-old male guinea pig rubs his perineal region on the cage sides. His anogenital region is packed with a thick, oily material mixed with hair and wood chip bedding. The area is gently cleaned as shown (58).
a What abnormalities are present?
b What is the anatomical term for this 'pouch'?
c What is the common name of the material that accumulates in this region and how does is develop?
d How would you treat the condition?

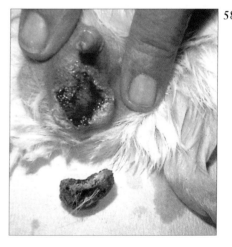

58

57 a The epididymal fat pad. The epididymal fat pad within the abdomen extends down into the inguinal canal when the testicle is in the scrotum. This large fat deposit prevents viscera from passing through the inguinal canal.

b With the rabbit in dorsal recumbency, clip and prepare the fur around the cranial aspect of the scrotum and penis as well as along the inner thighs for aseptic surgery. The first technique described is for a closed castration. Make a 1–1.5 cm incision through the scrotum longitudinally on each side of the midline about midway along the length of the scrotum. Grasp the tunic and remove the testicle from the scrotum with the tunic intact. The tunic tightly adheres to the end of the scrotum by the proper ligament of the testis. Break down this ligament to allow exteriorization of the testicle. Apply caudal traction to the testicle and use dry gauze to strip the facial attachments allowing the narrow portion of the cord to be exteriorized. Once the testicle has been adequately exteriorized, ligate the base of the tunic containing the cord using a two- or three-clamp technique. The disadvantage of this technique is that if the suture is not tight enough, the spermatic vessels may slip out of the ligature.

A second technique involves performing an open castration and closing the inguinal ring. Make an incision as described above and incise the vaginal tunic to allow exteriorization of the testicle, spermatic cord and vascular supply. Double ligate the spermatic cord and remove the testicle. Trace the vascular pedicle craniad and identify the inguinal canal. Place a single interrupted suture across the inguinal canal being careful not to compress the blood vessels passing through the canal.

A third technique involves an open castration removing only the testicle, leaving the epididymal fat pad intact. The fat pad prevents herniation of intestine through the inguinal ring. With all of these techniques, the scrotal incision may be left open to heal by second intention or sutured closed using either an intradermal pattern, tissue adhesive or skin staples. Use 4–0 absorbable synthetic sutures.

58 a The tissue is inflamed and ulcerated due to chronic irritation and superficial infection.

b The perineal sac. It is present in both sexes of guinea pigs. The anus is located at the very dorsal portion of the sac. In females, the vaginal orifice is separated from the anus by the perineal sac.

c 'Smegma' (even in females). It develops from perineal sebaceous gland secretions from the skin within the pouch. The pouch usually fills with trapped hairs, skin debris and oily secretions. The guinea pig does not have true anal glands. Build-up of material in this sac may be a sign that systemic disease is present because, normally, the guinea pig should be grooming this area. Grooming will decrease during systemic disease. The material is also present with low fiber diets causing chronic soft stools to accumulate. Two common systemic conditions include hypovitaminosis C and renal disease.

d Topical treatment is usually sufficient. Gently clean the area of debris and flush with saline once or twice a day. Application of a topical ointment aids in the removal of debris. Give a systemic antibiotic only if there is another disease present that warrants its administration. Administer vitamin C if necessary. Investigate and correct underlying disease.

59 Any condition in hamsters that results in a wet perineal area is often referred to as 'wet tail' by pet owners. It is not uncommon for these animals to be given a variety of over-the-counter medications.
a What etiologies should you consider when a hamster presents with a soiled, wet perineum?
b What is a common sequela to true 'wet tail'?
c How would you treat true 'wet tail'

60 Sudden deaths have occurred in a group of orphan European hedgehogs of mixed age. Before the deaths there was evidence of a green diarrhea flecked with blood. Bacteriological culture of the diarrhea was unremarkable. Two of the dead hedgehogs had rectal prolapse and the postmortem examination showed congestion of the small and large intestinal mucosa and focal pneumonia of the lungs. Histology of the large

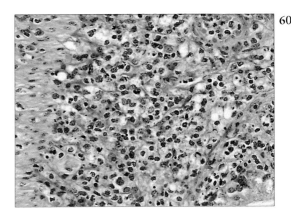

intestine showed a severe enteritis with marked thickening of the submucosa caused by infiltration of large mononuclear cells and neutrophils (**60**) (photograph courtesy of R. Munro). Numerous Gram-positive rods are present within the cells in the deep submucosa.
a What is the most likely cause of this syndrome?
b What are your differential diagnoses?
c How would you confirm a diagnosis?
d How would you manage the remainder of the group?
e Are there zoonotic implications?

61 It is common for male potbellied pigs to be castrated. At puberty, approximately three months of age, boars tend to become aggressive and may develop a strong musky odor. Accordingly, perform castration prior to this age. What technique would you use for castrating a potbellied pig boar?

59–61: Answers

59 a The most common cause of perineal staining is diarrhea from bacterial enteritis. Debilitated animals or those with overgrown incisors may have soiled perineums because they cannot groom themselves. An older hamster with polyuria caused by renal disease or endocrinopathies may have a urine-soaked perineum. Damp and dirty bedding can cause the fur to become wet and matted. Cystitis or cystic calculi can result in urine scald. Uterine neoplasia or infection can cause a mucoid, foul-smelling discharge that soils the perineum.

Classic 'wet tail' is proliferative ileitis resulting in a watery or mucoid diarrhea. It is primarily a disease of recently weaned hamsters. In the laboratory this disease can be created by infecting hamsters with chlamydial or campylobacter-like organisms. Conditions that can predispose a hamster to this disease include overcrowding and shipping stress. The mature single household pet is not likely to develop this disease. Enteritis in hamsters can also be caused by the use of antibiotics inappropriate for hamsters. Potentially fatal enterotoxemia caused by *Clostridia* spp. results when the antibiotic destroys the delicate balance of bacterial flora in the GI tract.

b Intussusception and rectal prolapse. The hamster often chews the prolapsed tissue and dies from secondary complications.

c Determine the cause of bacterial enteritis and correct it. Stop unsafe antibiotics, clean up the environment and isolate the patient. Give warmed subcutaneous fluids taking care to avoid the cheek pouches. Administer oral electrolyte solutions along with syringe feedings of a bland diet. Use enrofloxacin (10 mg/kg PO q 12 hours) or trimethoprim/sulfadimethoxine (30 mg/kg PO q 12 hours) as safe broad-spectrum antibiotics. Perform a fecal examination for parasites.

60 a *Salmonella enteritidis.*

b Screening of fecal samples and wild casualties reveal this as the predominant serotype, but *Salmonella typhimurium* is also occasionally isolated. There are reports of *Klebsiella* spp. being isolated from similar disease outbreaks. Other differential diagnoses include coccidiosis and heavy intestinal parasite infestation.

c Culture from the septicemic carcasses is likely to prove more rewarding than fecal sampling.

d Isolate the group and identify all stress factors and eliminate them if possible. Administer an appropriate antibiotic based on culture results to all the animals in the group. The use of probiotics is controversial, but is not likely to do any harm. Fluid therapy and nutritional support of the more severely affected hedgehogs are essential.

e Humans can develop salmonellosis and this potential should be carefully weighed when considering treatment of affected animals.

61 Perform castration in a manner similar to that done in a tomcat. Make a full-thickness incision through the scrotum over each testicle, exteriorize the testicle, open the tunic and ligate the spermatic cord with a strong (3–0) monofilament absorbable synthetic suture. Leave the scrotal incision open to heal.

Another technique involves a prescrotal incision as is performed in the dog. Palpate each testicle and gently push it towards the median raphe just cranial to the scrotum. Make an incision through the raphe and exteriorize the testicle, open the tunic and ligate the spermatic cord as previously described. Repeat the procedure for the opposite side using the same incision. Use a 3–0 monofilament, absorbable suture to close the subcutaneous and then subcuticular layers.

62 A five-year-old male rabbit acutely develops a head tilt. The external ear canal appears normal; however, the tympanum is opaque and bulging slightly. A spinal needle is inserted through the tympanum and an aspirate is removed for a microbiologic culture. Skull radiographs are taken and otitis media of the right ear is evidenced by the increased radiographic opacity within the osseous bulla.
a What is the name of the surgical procedure used to treat this condition?
b How would you carry out the procedure?

63 A pet African hedgehog has heavy flaking and crusting on the skin and some loss of quills (63).
a What is your diagnosis?
b What are your differential diagnoses?
c How could you confirm the diagnosis?
d How would you treat the condition?

64 Several rats from a large research facility have developed the ocular lesion seen in the photograph (64). A technician working with these rats has been anesthetizing them for two hours daily for a skeletal loading experiment, using a standard xylazine/ketamine anesthetic.
a What is your diagnosis?
b How would you manage this condition?

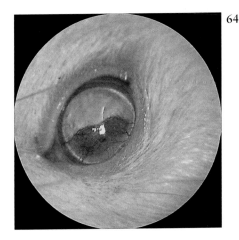

62 a Treat the infection aggressively with a ventral bulla osteotomy. This allows removal of infected tissue and debris. A microbiologic culture can also be obtained during the procedure.

b Palpate the bulla caudal and medial to the vertical ramus of the mandible. Make an incision about 4–5 cm long medial to the mandibular salivary gland between the angular process of the mandible and the wings of the atlas. Incise the subcutaneous muscle longitudinally along the same plane as the skin incision. Bluntly separate the digastricus muscle from the hyoglossal and styloglossal muscles, avoiding the hypoglossal nerve which runs on the lateral aspect of the hyoglossal muscle. Continue blunt dissection down (dorsally). Palpate the bulla as a round structure between the jugular process of the skull and the angular process of the mandible. Continue blunt dissection until the surface of the bulla is reached. Place a self-retaining retractor to maintain exposure. Incise the periosteum of the bulla with a scalpel and use a periosteal elevator to expose the bone of the osseous bulla. A small muscle courses along the ventral surface of the bulla from caudal to cranial. It has a tendinous attachment to the jugular process of the skull. Make the incision medial to this structure. The bone lateral to it is very hard and difficult to penetrate. Medial to this muscle, the bone is very thin and easily penetrated. Penetrate the bulla with a Steinmann pin and enlarge the opening with rongeurs. Collect samples for cytology and microbiologic culture. Irrigate the bulla and remove all debris. Remove the entire epithelial lining of the bulla using a small curette. During currettage, carefully avoid the dorsomedial aspect of the bulla cavity which is the location of the ossicles and the promontory. Damage to these structures results in vestibular signs.

Once the bulla has been adequately debrided, place an ingress-egress drain. Loosely close the muscles and subcutaneous tissues and routinely close the skin. Use the drain to irrigate the middle ear for 7–10 days. Maintain the rabbit on appropriate antibiotic therapy based on the results of the microbiologic culture and sensitivity. Take radiographs one month after treatment to assess the status of the disease.

63 a An infestation of skin mites, most often of the genus *Chorioptes*.
b Other ectoparasites, dermatomycosis, bacterial folliculitis and immune-mediated diseases.
c Mites are demonstrated on a microscopic examination of skin scrapings from the affected area.
d The simplest and most effective treatment is ivermectin (0.2–0.4 mg/kg PO or by SC injection) given as a single dose. Repeat the dose at two to three week intervals for a minimum of three treatments. Insecticidal whole body dips are not recommended due to the potential for toxicity. Thoroughly disinfect the environment since mites can survive off the pet for several days.

64 a This vascularized area of corneal ulceration in the central interpalpebral zone of the cornea occurs because the animals have their eyes wide open during anesthesia with these agents. Dehydration of the cornea plays a major role in the development of this lesion.
b Solve the problem by taping the lids closed during surgery. Consider protecting the eyes in this manner during any long procedure under anesthetic where ocular drying and subsequent trauma may occur if the eyes remain open. Alternatively, a sterile ophthalmic ointment may be placed in the eyes to keep them lubricated.

65 Endotracheal intubation in guinea pigs is very challenging. This shows the caudal aspect of the guinea pig oral pharynx (65a) (photograph courtesy of S. Jahn).

a What is the central circular opening called and what are its implications concerning anesthesia in this species?

b If placement of an endotracheal tube is required in a guinea pig, what obstacles must you overcome?

c If needed, how would you accomplish endotracheal intubation in a guinea pig?

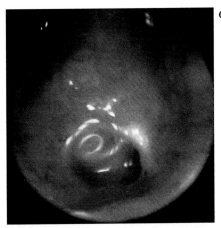
65a

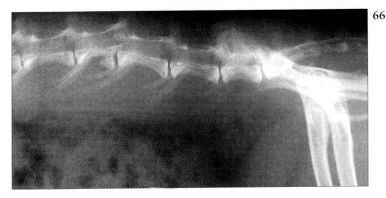
66

66 A three-year-old rabbit is acutely paralyzed in the rear limbs. The paralysis occurred after the rabbit was handled by several small children. The patient is alert and responsive, with normal mentation and normal appetite. The front limbs have no neurological abnormalities; the rear limbs are bilaterally hyporeflexive, and show no evidence of voluntary motion. This spinal radiograph was taken (66).

a What lesion is depicted in the radiograph?

b Why is this lesion commonly found in rabbits?

c What is the prognosis for this rabbit?

65, 66: Answers

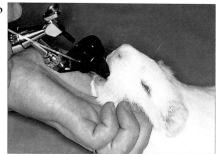
65b

65 a This opening is the palatal ostium. In the guinea pig the soft palate extends down to the base of the tongue. The small palatal ostium is the only opening between the oral pharynx and the more proximal aspects of the pharynx. It is difficult to pass either a stomach tube or an endotracheal tube through this small opening. Trauma to the soft palate can cause bleeding and subsequent asphyxiation of the animal.

b The mouth of the guinea pig is long and narrow and does not open very wide. Achieve access to the glottis by passing through the small palatal ostium. Prominent premolars and molars limit the space available for placement of a laryngoscope blade. The tongue is large, only the tip is freely movable, and food material often accumulates at the base. This can obstruct visualization of the palatal ostium and result in aspiration and airway obstruction. To overcome this problem, fast normal adult guinea pigs no more than two to four hours before anesthesia. Do not fast pregnant guinea pigs close to term as fasting can induce ketosis. If a guinea pig cannot be fasted before anesthesia, remove the food at the back of the mouth with cotton swabs after anesthetic induction.

c A guinea pig may be satisfactorily maintained under inhalant anesthesia using a small mask and appropriate gas flows. Endotracheal intubation is necessary when ventilatory support and control of the airway is required. A number of different techniques are described for endotracheal intubation of the guinea pig. Specially taper small laryngoscope blades (i.e. size 0 Miller blade) to a narrow point to facilitate visualization of the glottis. In another technique, use a lighted otoscope with a 3–4 mm cone. Place the anesthetized guinea pig in sternal recumbency. Introduce the lighted otoscope cone into the mouth, pass it over the tongue and use it to hold the ostium open allowing visualization of the glottis. Introduce a 3.5 Fr urinary catheter into the larynx (65b) (photograph courtesy of S. Jahn). Withdraw the otoscope cone leaving the catheter in place. Advance the endotracheal tube over the catheter and remove the stylet. Endotracheal tubes used in guinea pigs include 1.5–2.5 mm Cole or straight tubes. Modified intravenous catheters (14 gauge) or other atraumatic tubing are also used.

66 a A comminuted fracture is visible in the fourth lumbar vertebrae, resulting in misalignment of the vertebral column caudal to the fracture.

b Rabbits have extremely powerful rear limbs. If adequate control of the rear limbs is not maintained during restraint, vertebral fractures or luxations can occur if the rabbit kicks and hyperextends the spine.

c Poor, especially in cases such as this where alignment of the vertebral column is not maintained. Routinely euthanizing all rabbits with spinal fractures is not recommended as some rabbits can exhibit improvement weeks to months later with supportive care including cage rest and anti-inflammatory drugs. Each case should be individually evaluated.

67 Injectable anesthetics are often used to immobilize non-human primates.
a What is the intramuscular drug of choice for anesthesia in primates?
b What dosages are used in a great ape, spider monkey and marmoset?

68 A domestic ferret is proving difficult to handle.
a How would you restrain the ferret?
b How would you administer enteral and parental medications?

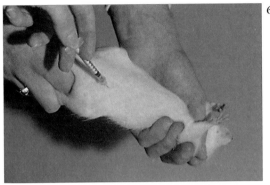

69

69 Injectable anesthetics are often administered to small mammals by the intraperitoneal route because it is a quick, simple and relatively pain-free method of drug administration (**69**). Are there any disadvantages to the intraperitoneal route?

67 a Ketamine hydrochloride. Tiletamine/zolazepam is also very effective, but the recovery time may be prolonged and undesirable.
b Administer ketamine (4–6 mg/kg IM) for a great ape, (10–20 mg/kg IM) for a spider monkey and (30–40 mg/kg IM) for marmosets and tamarins. Generally, the smaller the primate, the higher the dosage.

68 a It is rarely necessary to use gloves or other hand protection when handling pet ferrets. Kits under four months of age tend to 'play bite' more frequently than adults. Ferrets have a very keen sense of smell and some odors on the hand, especially food or other ferret body odors, can precipitate a bite. Place a drop of isopropyl alcohol on the gums of the biting animal to cause an immediate release. To examine the dorsum, rest the ferret's body along the forearm. To inspect the ventrum, place the patient on its dorsum by cupping its head in the palm of your hand and secure the ferret between your forearm and body. A ferret can also be 'scruffed', which calms down some active patients and allows better access for palpation of the cranial abdomen (68). Grasp the ferret by the loose skin along the back of the neck and suspend it over a table. Gently stroke downwards on its abdomen to aid relaxation. Use this hold for administering subcutaneous injections, cleaning the ears and trimming the nails. Ferrets love sweet treats and these can be employed to handle a fractious patient. Put a sticky sweet substance, such as a cat hairball laxative, on a tongue depressor and allow the ferret to lick it when weighing the patient or giving injections. For nail trimming without assistance, set the ferret in a sitting position in the lap, place a small amount of a sticky sweet substance on its lower abdomen and show the ferret the treat. While the ferret is licking the treat, trim the nails on all four paws. The ferret's abdomen should normally be relaxed and easy to palpate. The spleen, kidneys, stomach and bladder are all accessible.
b The easiest way to medicate a ferret orally is to mask the taste with a sweet or oily substance. Substances high in sugar content are contraindicated if the ferret is diagnosed with insulinoma. Sweet substances can be diluted with water and still maintain the taste desired. Give subcutaneous injections in the neck area as the animal is scruffed or along the back or shoulders. Administer intramuscular injections in the lumbar musculature or quadriceps.

69 Intraperitoneal administration of anesthetics has several disadvantages. Inadvertent injection into the viscera results in a failure of anesthesia induction. Intraperitoneal administration of irritant agents causes pain and an increased likelihood of peritonitis postoperatively. The inability to titrate the anesthetic dose is the most serious disadvantage. Since considerable variation in sensitivity to different anesthetic agents occurs, select an agent with a wide safety margin. If underdosage occurs, a 'top up' injection can be given intraperitoneal but this can be hazardous, so it is preferable to deepen anesthesia using a low concentration of an inhalational agent (e.g. 0.5–1% isoflurane).

70 A young doe in a herd of breeding rabbits has vesicles and ulcerated epidermis with a keratin rim on the vulva (70). There were no other clinical signs of disease in any of the other rabbits.
a What is your diagnosis?
b How would you confirm the diagnosis?
c How would you treat the disease?
d How would you control the disease?

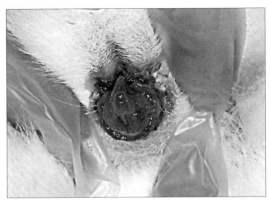

71 A chinchilla has diarrhea for several days after purchase from a pet store. There is now a rectal prolapse. These organisms measuring 15 μm in length are found on zinc sulfate flotation of the feces (71).
a What is this organism and is it causing the diarrhea?
b How would you treat the chinchilla?
c Are there any zoonotic considerations?

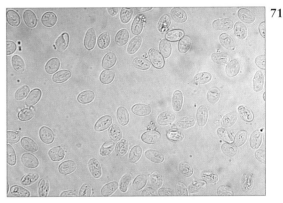

72 An adult rabbit has a facial draining nodule of two months duration (72) (photograph courtesy of R. Harvey). Previous administration of enrofloxacin for three days was unsuccessful. The rabbit lives outdoors in a wooden hutch.
a What is the presenting lesion?
b Which organism is most likely to be isolated from this lesion?
c What underlying disease factors may be important in the pathogenesis of this condition?
d What is your prognosis for this rabbit and how would you manage the condition?

70–72: Answers

70 Ulcerative lesions at the mucocutaneous junctions of the genitalia, anus, nose, eyelids or lips are typical of rabbit syphilis caused by the spirochete *Treponema cuniculi*. The infection is transmitted by direct contact between breeding rabbits and from mother to young. Mild lesions may resolve spontaneously, but infected rabbits become carriers with the spirochetes remaining latent in the lymph glands. Treponematosis is endemic and subclinical in many rabbitries. Other signs of this disease include abortions, metritis and infertility.
b Examine material scraped from the lesions under dark-field microscopy to demonstrate the spiral-shaped organisms. Diagnosis is confirmed by the direct immunofluorescent test. Use serological assays, the indirect IFA and a microhemagglutination test to detect subclinical carriers of the disease.
c Treat with three injections of benzathine penicillin G (42,000 iu/kg IM) given at seven-day intervals. The treatment results in regression of lesions and eliminates the infection. Although injectible penicillin is relatively safe, establish the rabbit on a high fiber diet to avoid potential cecal dysbiosis. Do not use beta lactam antibiotics orally in the rabbit.
d To control the infection in a herd, treat the breeders. To eliminate treponematosis in a herd, breed only seronegative rabbits.

71 a This is a cyst of a *Giardia* sp. *Giardia* can may produce diarrhea in many species of animals, including chinchillas. Identify it by size, the oval shape and the presence of nuclei, median bodies and flagella within the cyst. Zinc sulfate flotation is the test of choice to recover *Giardia* sp. from the feces. A direct fecal smear is usually insufficient and multiple very fresh samples need to be evaluated.
b Fenbendazole (25–50 mg/kg PO for 3–5 days) is probably the drug of choice for treatment, although metronidazole has also been used.
c The zoonotic implication of *Giardia* in chinchillas is unknown. It is best to be safe and advise clients that there is a slight zoonotic risk.

72 a A subcutaneous abscess.
b The organism most often isolated from such lesions is *Pasteurella multocida*, although *Staphylococcus aureus*, *Fusobacterium* spp., *Pseudomonas aeruginosa*, *Streptococcus* spp., *Corynebacterium pyogenes*, *Escherichia coli* and *Klebsiella* spp. have been isolated from similar lesions.
c Subcutaneous abscesses have a variety of causes including bite wounds or other skin trauma, periodontal disease, nasolacrimal duct infection, penetrating cheek wound and seeding from a distant infection.
d The prognosis in rabbits with subcutaneous abscesses is guarded. Perform diagnostic tests as needed to identify other disease that may alter the immune response. Obtain a bacterial culture and sensitivity. Culture the wall of the abscess rather than taking a sample of the discharge. Conventional therapy consists of aggressive surgery to remove as much of the abscess as possible. Remove affected teeth. Administer long-term antibiotic therapy. In cases where the entire abscess is not removed, keep the surgical site open and flush daily with an antiseptic solution to promote healing by granulation. Lancing and draining the abscess often results in recurrence. Some abscesses will recur at the same or different sites despite aggressive therapy. Other treatment modalities to use concurrently with surgery, such as injecting gentocin into the abscess wall or placing antibiotic-laden material into the abscess lumen, may offer promise.

73 This patas monkey is housed in an outdoor facility that meets or exceeds standards of primate housing specified in the regulations of the Animal Welfare Act (AWA), United States, 1991. Although AWA guidelines serve as a benchmark for primate facilities worldwide, practitioners should consult the regulations and legislation of their country.
a What are the minimum space requirements for non-human primates as specified in the AWA?
b What materials are generally recommended for cage construction and cage accessories?
c What disinfectants are recommended for cleaning cages, dishes and toys?
d What items in (73) fulfill the requirements of environmental enhancement to promote psychological well-being for this monkey?

74 A one year-old, neutered male ferret is experiencing chronic upper respiratory disease. It sneezes, coughs, is lethargic and has had a low-grade fever of 39.4°C for five to seven days. These signs have occured at least once a month for the last four months. The ferret appears to return to normal with the use of supplemental feedings and antibiotics. Amoxicillin and enrofloxacin have been used. The two other ferrets in the household appear to be normal. This one receives a high-quality commercial ferret pellet and is housed in a spacious cage. No other clinical signs are evident and its body weight appears normal.
a What are the differential diagnoses for this ferret?
b What further investigations would you conduct?

73, 74: Answers

73 a The minimum space provided to each non-human primate whether housed individually or with other nonhuman primates is determined by the typical weight of animals of its species and is calculated using the formula listed below. Brachiating species and Great Apes weighing over 50 kg must be provided with additional volumes of space in excess of that required for Group 6 animals to allow for normal postural adjustments.

Group	Weight (kg)	Floor area/ animal (m²)	Height (cm)
1	Under 1	0.15	50.8
2	1–3	0.28	76.2
3	3–10	0.40	76.2
4	10–15	0.56	81.28
5	15–25	0.74	91.44
6	Over 25	2.33	213.66

Examples of non-human primates included in each weight group:
Group 1: marmosets, tamarins, infants (less than 6 months of various species).
Group 2: capuchins, squirrel monkeys, juveniles (6 months to 3 years of age of various species).
Group 3: female macaques and African species.
Group 4: male macaques and large African species.
Group 5: baboons, non-brachiating species larger than 15 kg.
Group 6: Great Apes over 25 kg.
b Use stainless steel to construct cages. Build cage accessories such as perches, nest boxes, feeder boxes and various swings and platforms from non-porous materials that are durable and easily cleaned and disinfected. These materials include stainless steel and PVC piping.
c Clean cages, dishes and toys with disinfectants such as Roccal-D (Winthrop, New York, NY) or One-Stroke Environ (Ceva Laboratories, Overland Park, KS). The use of a 1:10 bleach solution may corrode surfaces.
d Items in this cage include a paint roller, a hard rubber toy, a ball, a hoop, plastic dishes, wire puzzle feeder device and a PVC climbing structure.

74 a Lymphoma, viral URI (including human influenza), bacterial URI (including *Bordetella* sp.), fungal disease and inhaled irritants. Intrathoracic masses can result in respiratory signs, such as dyspnea, coughing or wheezing. Cardiac disease can occasionally cause coughing, but it is rare in young ferrets and would not likely resolve with the use of antibiotics.
b The two most important diagnostic tests to perform initially are hematology and a thoracic radiograph. Additional diagnostic tests include thoracic ultrasonography, tracheal wash with cytology and culture and serum biochemistries. Take a detailed history to rule out environmental irritants. Substances that affect ferrets include clay cat litter dust, perfumed fabric softners or detergents used on bedding, carpet deodorant powders, aromatic bedding materials (pine or cedar chips) and aerosol room deodorants.

In this case thoracic radiography and ultrasonography was normal and the results of hematology are listed in question 75. There is no history of environmental irritants.

75 The hematology in question **74** revealed a WBC of 15,000 with 50% lympho-cytes, 45% neutrophils, 3% eosinophils, 2% basophils, PCV 45% and RBC 6.2 million.
a What is your diagnosis?
b What would be your next diagnostic step?

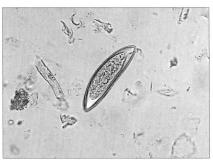

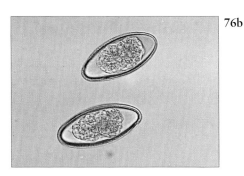

76 A breeder that sells 'germ-free mice' is receiving complaints from researchers that the mice are infected with worms. The company assures the researchers that fecal flotations performed weekly are negative for eggs, larvae, cysts and oocysts. The company maintains that the mice are negative when they leave the facility and become infected when they arrive at the research facility. One of the researchers performs a 'tape test' on a newly arrived group of mice and recovers eggs measuring 118–153 × 33–55 μm (**76a**). On sugar flotation, a second kind of egg is found that measures 89–93 × 36–42 μm (**76b**). The mice are 28 days old when they arrive at the research facility.
a What are the eggs?
b How is a 'tape test' performed and why is it necessary?
c Why are eggs found on the flotation performed at the research institution but not at the breeder?

75 a Lymphoma. The WBC count of 15,000 is greatly elevated over the normal range of 3,000–7,500 and the absolute lymphocyte count of 7,500 is also elevated over the normal high of 3,500. (It should be noted that when lymphocytes compose 60% or more of the differential diagnosis, this is also considered a possible sign of lymphoma regardless of the absolute lymphocyte count.) If the leukophilia was due to a bacterial infection, there should be a neutrophilia of 80% or higher present. Ferrets do not routinely have significant hemogram changes under the stress of examination or restraint. Although a definitive diagnosis of lymphoma cannot be made from a CBC, there is a strong suspicion due to the chronic nature of the illness and the extremely high absolute lymphocyte count. It is not unusual for immunocompromised lymphoma-affected animals to experience chronic recurring illness, such as respiratory or GI disease. These infections may respond temporarily to antibiotics or other supportive care only to recur days to weeks later.

b Obtain a lymph node biopsy or a bone marrow aspirate. Although some cases of lymphoma will have obvious local or generalized lymph node enlargement, this is not always the case. Where there is no apparent lymph node enlargement, any peripheral node may be used for biopsy. The popliteal or inguinal lymph nodes are easy to access and have minimal blood supply. Remove the entire lymph node rather than attempting to aspirate it. In the healthy ferret, the peripheral lymph nodes are surrounded by a large amount of perinodal fat making needle aspiration difficult. Obtain a bone marrow aspirate from the femur or ilium. Bone marrow aspirates are frequently less useful for diagnostic purposes than a lymph node biopsy. The diagnosis in this case was malignant lymphoma based on histopathology of the popliteal lymph node. Chemotherapy is a viable option for many of these cases.

76 a Both figures show pinworm eggs. Figure **76a** shows the common mouse pinworm *Syphacia obvelata* and **76b** is of the mouse pinworm *Aspisculuris tetraptera*. Note the flattened side of the *Syphacia* egg.

b The tape test is necessary to find the eggs of many pinworms, including *Syphacia*, because the female worms do not deposit eggs in the intestine. *Syphacia* eggs are not usually recovered on fecal flotations. The female worm crawls through the anal opening and cements eggs to the perianal skin. Press the adhesive side of a clear piece of cellophane tape against the anus and perianal skin. Stick the tape to a slide for microscopic examination. *Syphacia* adults live in the colon. The life cycle is direct and the larvated eggs deposited by the female are the infective stage. The prepatent period of *Syphacia* is 8–15 days.

c *Aspiculuris* adults live in the anterior colon and cecum where females deposit eggs rather than attach them to the perianal skin. Eggs, therefore, are present on flotation. Like *Syphacia*, the life-cycle is direct and larvated eggs are infective. The prepatent period, however, is approximately 23 days and eggs larvate to the infective stage in approximately six days. The breeding facility was performing fecals on the young mice and the worms were not old enough to produce eggs. In addition, the breeding facility was not performing the tape test.

77 The chinchilla is a true herbivore with typical rodent incisor dentition.

a What is the dental formula for the chinchilla?

b Is the incisor occlusion illustrated normal for the chinchilla (77a)?

c Describe the radiographic views most suited to demonstrating the occlusal planes in the chinchilla.

d What are the main features of interest when examining these radiographs?

e Why is epiphora a common indication of dental disease in rabbits and herbivorous rodents?

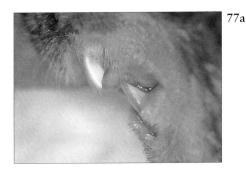

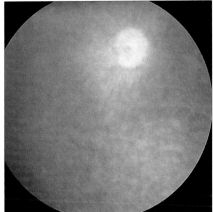

78 A juvenile Virginia opossum and a 13-lined ground squirrel are both hit by a car. Once stable, an ophthalmic examination is conducted. The photographs represent what is observed during the examination of the posterior segment of the eye of the opossum (78a) and squirrel (78b).

a What is your diagnosis for the opossum?

b What is your diagnosis for the squirrel?

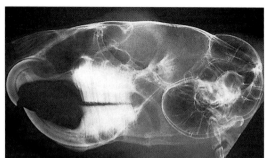

77b

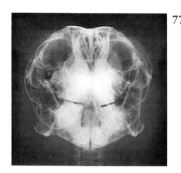

77c

77 a I$\frac{1}{1}$ C$\frac{0}{0}$ P$\frac{1}{1}$ M$\frac{3}{3}$

b Yes. Unlike rabbits, the majority of rodents maintain their jaws in a retrognathic position when at rest. In this position the cheek teeth are usually in occlusion. When gnawing, the jaw is brought forward and then moved in a dorsoventral direction. As many rodents have a relatively large normal range of rostrocaudal jaw movement, they can compensate for even quite dramatic jaw length discrepancies.

c The occlusal plane is horizontal in chinchillas. It can be demonstrated on lateral and rostrocaudal radiographs.

d A distinctive radiographic feature of the chinchilla is the large tympanic bulla. The lateral radiograph in 77b clearly shows the curvature and extent of the incisor teeth and the cheek tooth occlusal planes. Even with superimposition of the two sides, the position and length of the cheek teeth can be identified. Note the normal radiolucent germinal areas at the root apices. The rostrocaudal view in 77c demonstrates the temporomandibular articulations.

e The nasolacrimal ducts, which pass next to the root apices of the premolars and incisors, frequently become obstructed when the roots of these aradicular hypsodont teeth elongate or develop other periapical pathology. In addition to primary eye disease, a purulent ocular discharge may originate from lacrimal duct infection secondary to either simple obstruction or a tooth root abscess.

78 a This is a normal opossum eye. The opossum tapetum is unusual as it is located within the retinal pigmented epithelial cells rather than in the choroid. The fundus is richly invested with blood vessels (holangiotic vascular pattern) and the vascular tuft (not to be confused with hemorrhage) on top of the darkly pigmented disc is normal for this species.

b This is a normal squirrel eye. The optic nerve of the squirrel is located far superior in the fundus and is compressed into a thin line that traverses the fundus in a horizontal direction. The optic nerve always represents a blind spot in the visual field of an animal. Thus the shape of the nerve and its location (placed superior thereby visualizing the ground) makes it unlikely that the form of a predator could fully occupy its blind spot. This diminishes the chance of a predator remaining undetected.

79 The rabbit has a fragile lumbar spine that can luxate or fracture easily if the animal is handled incorrectly. Describe several techniques for picking up a rabbit.

80 Note the black oval 'lesion' on the flank of this hamster (**80**).
a What is this structure?
b What other rodent species have similar structures and where are they located?
c What are the known functions of this structure?

81 An adult European hedgehog is found by the roadside. There is concern because both hind limbs remain visible below the tibia even though the hedgehog is in its protective curled posture. A radiograph taken without anesthesia fails to reveal anticipated hindlimb trauma (**81**).
a What further tests would you carry out on the conscious hedgehog?
b What is your diagnosis?
c How would you confirm this?
d What is the prognosis?

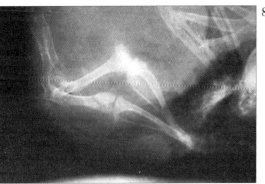

79 Always support the hind quarters of the rabbit when picking it up or transporting it. Remove rabbits from their containers on the floor to prevent potential 'escape' jumps from a high table. In addition, weigh the rabbit on the floor to prevent similar accidents. Three methods of picking up a rabbit are: (1) Grasp the loose skin over the back of the neck and lift it up while placing the other hand under the hind quarters (79). (2) Place an open hand under the thorax and, as the rabbit is lifted, grasp the thorax and use the other hand to support the hind quarters. (3) Grasp the thorax as in (2) with one hand and firmly grasp the lower lumbar area with the other hand. This firm grip on the lumbar area will frequently cause a fractious rabbit to stop kicking and it protects the area of the spine that is often luxated or fractured. In each of these cases the rabbit is transported immediately to the examination table where a mat or towel is placed to prevent the feet from slipping. Rabbits kicking out on a slippery surface can also damage their backs.

80 a This is a flank or scent gland. It is most prominent in mature male hamsters and becomes visible with hair thinning.
b Gerbils possess a prominent patch of sebaceous glands on the midventral abdomen; there are used for territorial marking and identifiying the pups. The gland is larger in the male. Guinea pigs have prominent sebaceous glands around the anus and over the rump, which produce waxy secretions used for territorial marking. The glands are common sites of cysts, abscesses, hyperplasia and neoplasia in these species.
c The flank gland serves as a means for olfactory marking of territory, individual identification and, possibly, sexual attraction.

81 a A neurological examination can be carried out. If the limb is fractured, pinching the web between the toes or a nail-bed may not produce the expected withdrawal reflex. However, in response to pain, the orbicularis muscle will contract, rolling the hedgehog into a tighter ball with the affected limb still exposed. Reaction of the panniculus carnosus muscle to this stimulation will cause reflex bristling of the spines. When there is no deep pain response to the toe pinch, test the neurologic integrity of the orbicularis muscle by gently pricking the skin over the muscle. This should produce the expected contraction. Lack of response to this series of tests indicates hind limb paraplegia. In this animal there was an absence of both the deep pain and the withdrawal response.
b Spinal cord trauma is the most likely diagnosis.
c Take lateral and ventrodorsal radiographs of the spine. Sedation is required for radiography except in very debilitated animals. Use either an injectable agent such as ketamine hydrochloride (20–80 mg/kg IM) or an inhalant such as isoflurane administered in an induction chamber or by face mask.
d If spinal damage is confirmed in a wild hedgehog, euthanasia should be considered due the extremely poor prognosis.

82 Routine tuberculin skin testing on all non-human primates minimizes the risk to the human caretakers from *Mycobacterium tuberculosis*.
a What is the recommended type of tuberculin, injection site and route of administration for the TB test?
b At what time interval is the test read and how frequently is testing performed?
c What constitutes a negative, a suspect and a positive test?

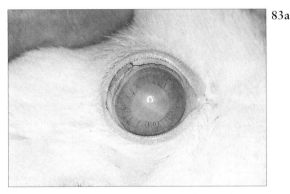

83a

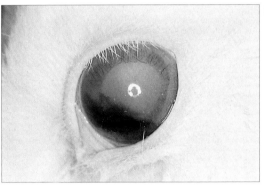

83b

83 An owner complains about chronically red eyes in a New Zealand White rabbit. The left eye has dramatically changed appearance in the past 24 hours (83a).
a What is the most likely cause for the clinical appearance of the eyes?
b What is present in the anterior chamber of the left eye (83b)?
c What is the appropriate term to describe this finding?
d What treatment would you recommend for this condition?

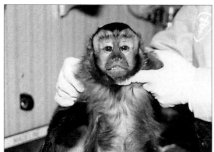

82 a Currently the standard is OT (mammalian tuberculin, Symbiotics, Inc. 1-800-247-1725) at 0.1 ml intradermally via a tuberculin syringe with a 25–27 gauge needle. Use as a primary injection site the eyelid and use the skin over the abdomen as a secondary site (**82a**).

b Read tuberculin tests at 24, 48 and 72 hours post injection. Tuberculin test non-human primates at least yearly. Many colonies and laboratory facilities test as often as every three months. Test humans exposed to non-human primates annually although many facilities require employee testing every six months.

c Consider a test negative if, at 48 and 72 hours, there is no edema or swelling at the injection site. Bruising and a small amount of erythema without any swelling still constitute a negative test. Regard a test as suspicious if, at 48 and 72 hours, there is a larger degree of erythema with minimal swelling or there is slight swelling without erythema. Read a test as positive if, at 48 and 72 hours, there is obvious swelling of the palpebrum with a drooping eyelid associated with erythema (**82b**). The eyelid may even be closed and necrotic. Repeat a suspicious test in 7–10 days in the other eyelid or in the abdominal skin. Isolate the animal from others until its status is completely determined.

83 a Lens induced uveitis. Both eyes have hypermature cataracts. The leakage of lens protein initiates an autoimmune response that results in chronic inflammation of the uveal tract.

b Blood.

c Hyphema. The severe inflammation in the left eye results in frank intraocular hemorrhage. The vessels of the chronically inflamed iris are so fragile that even minor trauma results in bleeding.

d Treat lens induced uveitis by controlling the inflammation. Use steroid and non-steroidal anti-inflammatory ophthalmic agents. In rabbits, use steroids with great caution. Removal of the cataract to control the inflammation is not practical for most rabbits. Once the eye becomes inflamed from lens induced uveitis, the likelihood of complications subsequent to cataract surgery increases dramatically.

84 At the conclusion of a surgical procedure to remove a mammary tumor in a rat, several swabs have been used to control hemorrhage and clear the surgical site of blood. Is this degree of blood loss likely to be significant?

5a

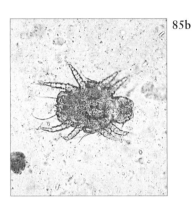

85b

85 This rabbit (85a) has developed dandruff and is losing hair. On physical examination, reddened, hairless, scaly patches are noted over the head and back. On a fecal flotation, the organism on the slide was recovered (85b).
a What is this organism?
b How would you treat the condition?
c How would you advise the owners?

86 All non-human primates are susceptible to tetanus and measles virus and the Great Ape species are also susceptible to poliomyelitis virus.
a What is the degree of susceptibility of non-human primates to each of these three diseases and what level of exposure do they need to contract the disease?
b What is the degree of severity of each of these diseases for non-human primates?
c What is the current recommended vaccination schedule for non-human primates?

84 The blood volume of rats is approximately 70 ml/kg, so the average adult female rat has a blood volume of 15–20 ml. Loss of more than 20% of blood volume usually causes signs of circulatory failure. A 200 g rat will only need to lose 4–5 ml of blood for this to occur, so even one blood-soaked swab represents significant hemorrhage. Avoid circulatory failure due to blood loss by careful surgical technique. Ligate blood vessels and use meticulous hemostasis. Cannulate the lateral tail vein with an over-the-needle catheter and infuse plasma volume expanders, balanced electrolyte solutions or even whole blood.

85 a *Cheyletiella parasitivorax*. This organism is referred to as 'walking dandruff' and its presence in the fecal examination is simply fortuitous as a result of grooming. Identify the mite by the large palpal claws on its anterior surface. It can often be seen grossly visible moving in the heavily scaled areas. Use clear cellophane tape to collect the mites from the fur and skin. Examine the tape microscopically.
b Administer ivermectin (0.20–1.0 mg/kg SC q 14 days) for two to four doses or treat the rabbit by dusting with carbaryl or permethrin powder at weekly intervals. Alternatively, use lime/sulfur dips every two weeks for two to three treatments.
c *C. parasitivorax* survives in the environment longer than most mites. The premises remain a source of infection and should also be treated. Flea elimination products, including dessicants that are safe for cats work well, but prolonged treatment may be necessary. Use these products cautiously, especially in debilitated or obese rabbits, as deaths have been reported. On occasion the mites affect humans and other animals.

86 a Non-human primates have a low to moderate susceptibility to tetanus and need a moderate exposure to become infected. Tetanus can be fatal even with prompt treatment. Primates are moderately susceptible to the measles virus and need a low to moderate exposure level to contract the disease. This is also dependent upon the level of human contact and the level of measles virus in the human population.
b Severity ranges from an inapparent infection to death. Great Apes are very susceptible to the poliomyelitis virus at low exposure levels. Severity of this disease ranges from inapparent to fatal.
c Recommended vaccination schedule for non-human primates:

Age	Type of vaccine
2 months	Tetanus[1]; trivalent oral poliovirus[2]
4 months	Tetanus[1]; trivalent oral poliovirus[2]
6 months	Tetanus[1]
15 months	Measles[2]
18 months	Tetanus[1]; trivalent oral poliovirus[2]
4–6 years	Tetanus[1]; trivalent oral poliovirus[2]
10–12 years	Measles[1]
14–16 years	Tetanus[1]; trivalent oral poliovirus[2]
Every 10 years	Tetanus[1]

[1] All species.
[2] Great Apes only, use current human pediatric poliovirus product/route recommendations.

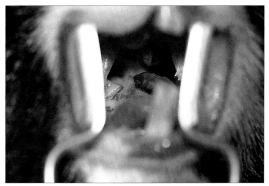

87 The guinea pig is a hystricomorph (porcupine-like) rodent and a true herbivore.
a Why is guinea pig enamel atypical for a rodent?
b The guinea pig has the same dental formula as the chinchilla. How does the guinea pig's cheek tooth dentition differ from that of the chinchilla?
c What effect does cheek tooth overgrowth have on the guinea pig's incisor occlusion?
d What is the problem shown in 87a and how would you treat it?

88 A three-year-old intact female rabbit develops an acute onset of generalized weakness, depression and incoordination (88). It is obese, maintains a splay-legged position, is weak on all four limbs and is reluctant to move. Spinal reflexes are normal in all four limbs. The breath has an acetone odor and several masses are palpable within the uterus. The rabbit is housed with an intact male and has had two normal litters previously, the most recent occurring one year ago.
a What is your diagnosis?
b How would you confirm your diagnosis?
c How is the condition treated?
d What is the prognosis?

87b

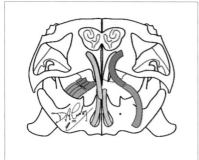

87 a Healthy rodents typically have yellow/orange pigmented enamel on their incisor teeth. Guinea pigs do not.

b There is a more significant rostrocaudal divergence of the guinea pig's dental arcades. Guinea pig cheek teeth are also markedly curved, compensating for a wider degree of anisognathism and resulting in the occlusal planes being angled at about 30° to the horizontal plane (**87b**). Additionally, the occlusal planes pass directly through the mandibular condyles and the tips of the incisor teeth in the guinea pig.

c Cheek tooth overgrowth tends to prevent the dental arcades from being brought fully into occlusion. It also prevents the mouth from closing fully in the normal resting jaw position. As a result, the mandible is held in a more prognathic position. This results in secondary malocclusion and overgrowth of the incisors.

d Bridging of the tongue by the mandibular premolars. The tooth on the left has bridged across the top of the tongue and is markedly discolored, while the overgrowth of the premolar on the right side is much less obvious. As the cheek teeth overgrow and the mandible is forced forwards, the mandibular premolars no longer occlude, resulting in overgrowth. This eventually leads to bridging over the tongue. Until this 'tongue tied' state develops, the guinea pig can usually cope without showing marked signs, so the problem is rarely detected early. Cut back the grossly overgrown teeth under anesthesia and reduce the crown height of all the teeth to approximate normal occlusion. Use analgesics and assist feeding after trimming until the patient is able to eat on its own. Examine these patients every two to three months because recurrence is common. Provide a diet high in grass hay and herbage to encourage frequent chewing and a natural level of tooth wear. Ensure adequate supplementation with vitamin C as deficiency affects collagen production and maintenance. Scurvy causes weakening of the periodontal ligament and may be a primary factor in cheek tooth malocclusion.

88 a Pregnancy toxemia is the most likely diagnosis. Although the cause of pregnancy toxemia in the rabbit is unknown, hepatic lipidosis is usually observed at necropsy, and the syndrome is seen most commonly in does which are obese and/or have fasted. Pregnancy toxemia usually occurs in the last week of gestation, although it is sometimes seen immediately postpartum. As in this case, affected does develop generalized muscle weakness leading to the splay leg appearance, anorexia, tachypnea and an acetone odor to the breath due to ketoacidosis.

b Radiograph the abdomen to confirm pregnancy. Use urinalysis to demonstrate ketonuria, proteinuria and an acid pH.

c Provide supportive therapy including IV or IO fluids and injectable antibiotics.

d Guarded to grave. If the doe does not respond to therapy and death is imminent, a Cesarean section should be performed if the owner wants to attempt to save the litter. Advise the owner that raising orphan rabbits is often unrewarding. Does used for breeding should not be allowed to become obese.

89 A ferret is diagnosed with a large spleen but does not appear clinically ill.
a What diagnostic tests would you use to determine the cause of the large spleen?
b Would you remove the spleen?

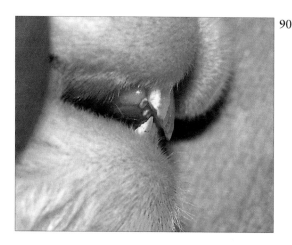

90

90 A mature pet rabbit is anorectic for two weeks, is reluctant to take syringe feedings and is losing weight. There is a history of recurrent inappetence which was previously managed by syringe feeding a slurry of liquidized rabbit pellets until the rabbit resumed eating. It was raised on a concentrate ration of pellets and grain without access to hay or fresh herbage. The rabbit has normal resting incisor occlusion, the tips of the mandibular incisors resting between the first and second maxillary incisors (90).
a What is the normal chewing action in the rabbit?
b What is the cause of inappetence in this case?
c What other signs would you associate with this problem?

89 a The diagnostic tests to determine the cause of the large spleen begin with the history and physical examination. If the history indicates a ferret in good health, extramedullary hematopoiesis may be a likely cause of splenomegaly, although some ferrets with lymphosarcoma may also have no clinical signs of disease. The physical examination of the ferret dictates the diagnostic steps to be undertaken. Spleno-megaly due to extramedullary hematopoiesis is characterized by an enlarged spleen with rounded borders. The organ palpates with a smooth surface and a spongy consistency. No other organ enlargement is present. Palpation of a 'lumpy' or 'hard' spleen is likely due to neoplasia. Hematology and serum biochemistry panels provide useful information. Alterations in the hemogram may be seen with neoplastic and infectious disease. Radiographs confirm the presence of splenomegaly. Use ultrasound to determine the splenic architecture. Extramedullary hematopoiesis does not change the normal architecture of spleen, whereas other disease, especially neoplasia, distorts the splenic architecture. Do a percutaneous splenic aspirate to determine the cause of splenomegaly. Scruff the ferret and place it on a flat surface while isolating the spleen with one hand. Aspirate the spleen with a 25 gauge needle attached to a 3 ml syringe. If neoplasia is present, it is usually diffuse and found on an aspirate. The ultimate diagnostic procedure is exploratory laparotomy and splenic biopsy.
b In most instances, remove the spleen if neoplasia is present. Successful treatment of lymphosarcoma may not depend on splenic removal. Typically, removal of the spleen for extramedullary hematopoiesis is not necessary. Infrequently, splenomegaly causes the splenic volume to be so great as to cause displacement of other abdominal organs. If this displacement is severe, it may cause the ferret to be uncomfortable and even partially anorexic. This is an indication for splenectomy even if the cause is extramedullary hematopoiesis.

90 a When eating natural foods, the lips are used for prehension and herbage is cut into manageable sections using the incisors. Rabbits use their incisor teeth in a lateral slicing action. The food is then crushed and ground using the cheek teeth prior to swallowing. The free end of the tongue is relatively thin and mobile, and is used to assist the lips with prehension and the transfer of food from the incisor to the cheek tooth region. The body of the tongue is fairly solid with a raised section between the cheek tooth arcades. It is used to move food between the grinding surfaces of the cheek teeth and form the ground material into a bolus for swallowing.
b The most likely cause is mucosal pain resulting from enamel spikes on one or more of the cheek teeth causing buccal or lingual irritation. A concentrate ration (grain and/or pellets) requires much less chewing for each unit of energy intake when compared with dry or fresh herbage, resulting in inadequate wear to the cheek teeth.
c With a concurrent reduction in the degree of lateral jaw excursion during chewing, the buccal surfaces of the maxillary (see **91a**) and the lingual surfaces of the mandibular cheek teeth (see **91b**) are worn least and sharp points or spikes of enamel form. Even very minor irregularities on the tooth surfaces may abrade or ulcerate the contacting lingual or buccal mucosa causing significant pain. This leads to inappetence, dysphagia, quidding (food dropping from the mouth during chewing) excessive salivation, reduced food intake and weight loss. Epiphora is occasionally present due to reflex lacrimation. Cheek abscesses can form secondary to mucosal penetration.

91 Shown are the buccal surfaces of the maxillary (**91a**) and the lingual surfaces of the mandibular cheek teeth (**91b**) of the rabbit in question **90**.
a How would you treat the rabbit's condition?
b What advice would you give to prevent this problem?

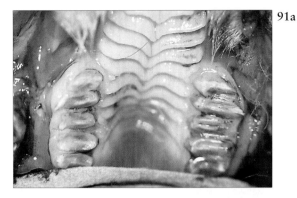

92 A monkey develops halitosis, anorexia and intermittent diarrhea. Oral examination reveals severe gingivitis, caries and periodontal disease (**92**).
a What are your differential diagnoses?
b What diagnostic tests would you recommend?
c How would you treat the condition?

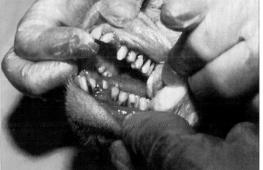

91, 92: Answers

91 a Textbooks generally recommend trimming or filing off sharp enamel spikes and restoring a normal occlusal surface. This is more easily said than done, particularly as the occlusal plane is not horizontal in rabbits. In many cases, there is an excessive curvature of the tooth roots leading to apparent tipping of the affected teeth out of alignment in the dental arcade. These teeth are subject to abnormal occlusal forces which may tip them further out of alignment. In early cases, an attempt at treatment is usually worthwhile provided that no complicating factors are found during pre-operative assessment. Examine these patients every two to three months because recurrence of enamel spikes is common.

In extreme cases, progressive damage to the tongue may lead to extensive scarring. Because it is only the mucosal surface that is sensitive, its destruction and scarring can allow the pain to subside until a new area becomes involved. In addition to removing the offending enamel spikes, provide supportive therapy, including analgesia, to get the animal through the acute stage of the disease. Re-examine the patient at regular intervals because some rabbits require life-long treatment of recurring enamel spikes. Improve the diet to include hay and fresh herbage.

b Feed rabbits a diet high in fresh grass hay and herbage. Such a diet requires thorough chewing and provides a natural level of tooth wear. Pellets and grain should only be provided for individuals with an exceptionally high energy demand, such as breeding does, debilitated animals and some fast-growing youngsters.

92 a Vitamin C deficiency, a diet high in sugars, a lack of dental prophylaxis, shigellosis, renal disease, septicemia, tooth root abscesses, *Candida* sp. infection, secondary bacterial infections due to immunosuppression, periodontal disease and gingivitis secondary to metabolic diseases (i.e. diabetes mellitus) and squamous cell carcinoma or other oral neoplasias. The differential diagnoses for the intermittent diarrhea include bacterial infections particularly *Campylobacter* sp., *Shigella* sp., *Salmonella* sp., *Clostridia* sp., Gram-negative bacterial overgrowth, parasitic infections such as *Entamoeba histolytica*, cryptosporidium, giardia, nematodes and cestodes, viral infections, lymphoplasmacytic or eosinophilic colitis, proliferative bowel disease, inflammatory or irritative bowel disease, GI tract neoplasia, irritated bowel syndrome, foreign body gastroenteritis and stress. Dietary causes of diarrhea include food allergies, lack of fiber or foods with laxative properties.

b Perform hematology, a serum biochemistry panel and viral serology. Culture the feces for aerobic and anaerobic bacteria and for *Salmonella* sp. and *Shigella* sp. Examine a direct fecal smear and a fecal floatation. Perform an ELISA test for *Giardia* sp. and *Cryptosporidia* sp. on the feces. Radiograph the teeth and skull to assess the degree of bone resorption around tooth roots. In addition, use radiographs to assess overall bone quality and the presence of lytic or proliferative bony lesions indicating the presence of abscesses, osteomyelitis or neoplasia.

c Until a positive diagnosis is made, treat the diarrhea symptomatically with bismuth subsalicylate (262 mg/3–5 kg body weight PO q 6–8 hours). Extract severely damaged or dead teeth although pulpotomy or root canal with endodontic therapy is preferred. Review diet and husbandry practices including sanitation and disposal of feces. Discuss any potential zoonotic diseases. Give additional supportive care such as softened foods, fluid therapy and analgesics for the dental pain. In this case, the definitive diagnosis was gingivitis due to shigellosis and the organism was sensitive to trimethoprim/sulfa.

93 A three-year-old neutered male ferret has a red, inflamed draining tract to the left of the anus. A foul smelling, purulent material is emanating from the wound.
a What is your diagnosis?
b How would you manage the condition initially?
c What surgery would you use to resolve the condition permanently?

94 An adult European hedgehog with erythema and pruritus of the haired areas of the skin is shown. There is some quill loss over the spine but the condition is most obvious on the face where the skin is thickened and crusted with dried serum (**94**).
a What has caused this skin condition?
b What laboratory test will confirm this?
c How would you treat the disease?
d Does this condition have significance for the hedgehog's future health?

95 A young boy who recently purchased a rat from a pet shop notes that both of its eyes appear cloudy (**95**). He is concerned that he has done something wrong in caring for the animal.
a What is your diagnosis?
b What advice would you give to the owner?

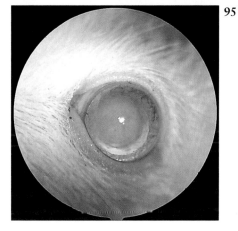

93 a A ruptured abscess of the left anal gland is the most likely diagnosis. Other possibilities include neoplasia of the glandular or muscular tissue of the area. This condition also occurs with incompletely removed anal glands resulting in fistulous tracts.

b Thoroughly flush the infected tract. Administer a broad-spectrum antibiotic, such as amoxicillin (15–30 mg/kg PO q 12 hours) or trimethoprim/sulfadiazine (30 mg/kg PO q 12 hours). Apply warm compresses to the perineum two to three times daily to reduce inflammation. When the inflammation has subsided in approximately 5–10 days, remove both the affected and normal anal glands. The other anal gland often becomes affected later if not removed.

c Under general anesthesia, isolate the anal duct opening and make a circular incision around it. Dissect the glandular tissue around the duct to a depth of 2–5 mm. After this, the gland may be bluntly dissected from the subcutaneous tissues and anal sphincter muscle. Either leave the small skin incision open or close it with a single, 5–0 absorbable suture or with tissue adhesive. In ferrets, these glands are relatively large and the contents have a strong unpleasant odor. Drop soiled sponges and the anal glands into a dilute bleach solution to neutralize the odor.

94 a A skin mite infestation which may have taken as little as three months to cause severe disease.

b Microscopic examination of a skin scraping from the affected areas will confirm the cause. Three species of mites are implicated in this disease. In Europe and New Zealand, the most common species found is the psoroptid mite *Caparina triplis*, whereas in the UK, *Sarcoptes* spp. and *Notoedres* spp. are more frequently isolated. Close examination with the naked eye or using a hand lens allows direct visualization of *Caparina* mites moving at the base of the spines. Although less than 0.5 mm across, they are white and contrast markedly with the dark skin of the hedgehog.

c Ivermectin (0.2–0.4 mg/kg SC) is the treatment of choice. Injections are repeated at two to three week intervals for up to three total treatments. It is important to clean the environment in the case of hedgehogs held in captivity.

d *Trichophyton erinacei* has been isolated from the feces of some skin mites and damage to the skin may predispose the hedgehog to this or other dermatologic disease. In addition, serum accumulation on the skin may attract flies and the risk of myasis. Studies in New Zealand concluded that hedgehogs with skin mite infestation were less likely to survive hibernation.

95 a There is a nuclear cataract in the lens. The eye appears otherwise normal.

b Reassure the owner that nuclear cataracts occur during the development of the animal *in utero* and that nothing he has done has caused the problem. A number of genes in laboratory rodents cause cataracts with various morphologies and this is probably one of them. Advise the owner that his pet will not suffer from its somewhat compromised vision. Even when completely blind, these rodents show few if any behavioral abnormalities. This rat still has considerable peripheral vision.

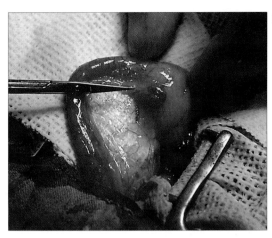

96

96 A seven-year-old ferret has numerous episodes of lethargy, ptyalism and collapse over a six-month period. Hematology and serum biochemistry values are normal except for a mild elevation in SGPT and a blood glucose of 2.8 mmol/l (32 mg/dl). An abdominal exploratory is performed. A mass is found at surgery (96).
a On what structure is this mass found?
b What is your diagnosis based on the history and the findings at surgery?
c How would treat the condition?

97 A seven-year-old pet rabbit is acutely anorectic. To aid in the diagnosis of the disease process, blood samples are collected.
a What are four sites that can be used for venipuncture in this rabbit?
b Describe methods of venipuncture in the rabbit.

96 a Pancreas.
b A pancreatic beta cell tumor or insulinoma. Hyperinsulinemia results in hypoglycemia leading to episodic weakness. In the ferret, ptyalism may be a sign of nausea.
c With surgery, medicine or a combination of both. Medical treatment includes frequent feedings, prednisone and diazoxide. Although surgery is the treatment of choice as it prolongs the ferret's life, it is not always possible. If the ferret has systemic disease (i.e. decompensated heart disease), if the owner cannot afford surgery, or if there are other extenuating cricumstances, then the ferret is maintained with medical therapy. Owners need to realize that eventually malignant beta cells will spread to other organs preventing glucohomeostasis with either medicine or surgery.

Perform an exploratory laparotomy with careful inspection of all the abdominal organs. Gently palpate the pancreas feeling for discrete, firm nodules. More than one insulinoma may be present in the pancreas The insulinoma can be as small as 1–2 mm in size. Occasionally, only diffuse pancreatic thickening may be present. Perform a nodulectomy in the body of the pancreas by shelling out the nodules. Do a partial pancreatectomy if tumors are located in the right or left limb. Use hemostatic clips for a partial pancreatectomy. Bleeding is minimal. Biopsy the liver and any abnormal lymph nodes. Perform a routine abdominal closure. For suture material, a 4–0 monofilament, synthetic absorbable can be used. Postsurgically, monitor the blood glucose concentration until the ferret is euglycemic. Infrequently, ferrets will need to be maintained on either prednisone and/or diazoxide (USA) in the immediate postsurgical period. Recheck the ferret 7–10 days after surgery and evaluate fasting blood glucose at that time. Because surgery is rarely curative, it will be necessary to monitor the fasting blood glucose at regular intervals for the rest of the ferret's life. At some point in time, it is likely that the ferret will need to be placed on medication to remain euglyclemic.

97 a There are a number of suitable sites for venipuncture in the rabbit. These include the jugular vein, lateral saphenous vein, cephalic vein and the auricular artery and veins.
b A butterfly catheter or a needle without a syringe attached is used to collect blood from the marginal ear vein or central ear artery. Use the ear vessels with caution. Rabbit veins are fragile and prone to hematomas. Damage to the ear vessel may lead to necrosis and loss of part of the pinna. Necrosis is also likely to occur if medications are administered through ear vessels. Rabbit skin is easily torn by clippers and it is safer to pluck hairs from around the venipuncture site and/or use alcohol liberally to wet down the area. Use a 25 gauge needle attached to a 1 ml syringe for the small-diameter veins, such as the cephalic and saphenous veins. For small rabbits, use an insulin syringe attached to a 27 gauge needle for venipuncture of the cephalic vein. Restrain a rabbit for cephalic venipuncture in a manner similar to that of a dog or a cat. The lateral saphenous is normally an easily accessible vein for venipuncture. The rabbit is held on its side and the rear leg is held off around the quadriceps. The lateral saphenous has numerous branches that can be utilized. Use a 22 gauge needle attached to a 3 ml syringe for the jugular vein. In dwarf breeds, a 25 gauge needle may be necessary. Hold the rabbit in sternal recumbency on an edge of a table with its forelegs pulled down and its head held up. The vein is in the jugular groove and is superficial. In obese rabbits, there may be fat over the jugular vein decreasing visualization.

98 Two adult rabbits are housed in a hutch in the garden of a semi-rural area in the UK. The previous year, both the rabbits were vaccinated with an attenuated Shope fibroma virus vaccine. Both rabbits have recently been exposed to a field strain of myxoma virus resulting in one rabbit's death. The surviving cagemate (98) developed lesions on its face seven to ten days after exposure, with no other clinical signs.
a What is your diagnosis?
b Are there any factors in the previous history of this rabbit which tends to strengthen your diagnosis?
c How would you confirm the diagnosis?
d How would you manage the conditon?
e What is your prognosis for this rabbit?

99 A mouse has been prepared for abdominal surgery (99). What are some likely clinical consequences?

100 Hypovitaminosis C is commonly seen as a cause of disease in pet guinea pigs.
a What diagnostic tests would be beneficial in diagnosing this condition?
b What is your therapeutic approach for a guinea pig with hypovitaminosis C?

98 a These lesions are likely to be the result of myxomatosis, caused by a pox virus.
b Myxomatosis is usually an acute infection with high morbidity and mortality, although less virulent strains of virus are recorded. In this case, it is likely that the rabbit was either exposed to a non-fatal infection in the previous season, or was vaccinated successfully with Shope fibroma virus (available in the UK). The degree of protective immunity produced by this vaccination in the UK is variable. The rabbit that died likely did not develop protection against myxomatosis despite previous vaccination.
c Confirm the diagnosis with biopsy and histopathology of the lesion or virus isolation. It is important to submit intact epidermis.
d Consider prophylactic treatment with an antibiotic such as enrofloxacin (10 mg/kg PO, SC or IM q 12 hours). No specific treatment is available for the viral infection, which is self-limiting
e The prognosis is good, as long as secondary bacterial respiratory disease is avoided.

99 Although maintenance of asepsis is important in rodents, removal of large areas of insulating fur and the use of excessive quantities of an alcohol-based skin-disinfectant increases the rate of heat loss. Hypothermia is a major problem when anesthetizing small rodents. This is due to their high surface area:body weight ratio resulting in more rapid losses than in larger species. When coupled with the depressant effects of anesthesia on thermoregulation, this results in a substantial fall in body temperature. To decrease the risk of hypothermia, every effort is made to minimize heat loss and supplemental heating is provided using heating blankets. Use surgical drapes, reduce the area of fur removal and minimize the quantity of skin disinfectant applied to help decrease the chance of hypothermia.

100 a The diagnosis is based primarily on the clinical signs and history. Radiographs may reveal enlarged costochondral junctions and abnormal epiphyseal growth centers in the limbs.
b Initially, supportive care is given and, secondly, the diet is supplemented with vitamin C. Use injectible ascorbic acid (50–100 mg/kg IM or SC q 12–24 hours). Switch to oral supplementation once the patient improves. Administer vitamin C orally (30–50 mg/kg q 12–24 hours) in liquid, capsule or tablet form. Provide fluid therapy and gavage feedings as needed until the patient can eat on its own.

When recovery is complete, discontinue oral vitamin C supplementation. Provide a diet rich in ascorbic acid by including half to one cup daily of fresh foods, such as dark green leafy vegetables (kale, mustard greens, dandelion greens, Swiss chard, cabbage). Alternatively, add vitamin C to the drinking water (100 mg/cup of water). The disadvantage to this method of supplementation is that the water should be changed at least once daily due to the rapidly decreasing potency of the vitamin when exposed to light and water. In addition, the bitter taste vitamin C can impart may decrease the guinea pig's water consumption. The guinea pig diet should also include unlimited fresh grass hay.

Because vitamin C is labile, do not depend on guinea pig pelleted food as the sole dietary source of this vitamin.

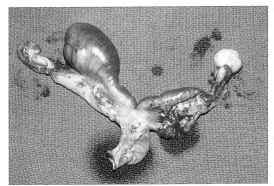

101

101 A six-year-old female rabbit has a bloody vaginal discharge for the last two months. The patient has resided as the only pet in the same household since eight weeks of age. On physical examination, a palpable caudal abdominal mass effect is noted and a hematocrit of 22% is found.
a What diagnostic tests would you now perform?
b What organ is shown (101)?
c What are your differential diagnoses for the enlarged uterus?
d Which is the most likely cause?
e What treatment plan would you recommend?

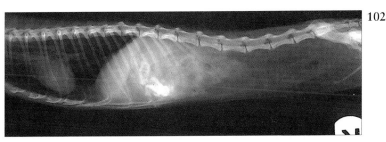

102

102 A lateral abdominal radiograph of a young ferret presented for lethargy, anorexia and diarrhea is shown (102).
a What is your diagnosis?
b What is the surgical procedure to treat this condition?
c What is the prognosis?

101 a Use radiographs to help define the location of the mass. If malignancy is part of the differential diagnosis, examine the lung and liver radiographically for metastasis. Use abdominal ultrasound to further define the location and architecture of the mass effect.
b An abnormally enlarged uterus.
c They include uterine adenocarcinoma, benign uterine hyperplasia, pyometra, cystic endometrial hyperplasia, uterine aneurysm and a uterus enlarged with fetuses.
d Uterine adenocarcinoma because it is a single pet with no history of breeding. Uterine adenocarcinoma is the most common reproductive disease of unspayed female rabbits over two years of age.
e If there is no evidence of metastasis, perform an ovariohysterectomy. At surgery, biopsy the mesenteric lymph nodes and liver to rule out metastasis. If there is no tumor spread, then the prognosis for a normal life span is excellent. If there is metastasis, then the prognosis is very poor. Radiograph and/or ultrasound the patient in three months and again at six months to look for metastasis. Rarely will metastasis occur after an ovariohysterectomy when no metastasis was seen grossly at surgery.

102 a Radiopaque foreign bodies in the small intestine.
b Perform abdominal exploratory surgery with careful palpation of the entire GI tract, including the stomach. This detects other foreign bodies not visible on the radiograph. If the foreign object is in the stomach, perform a routine gastrotomy. Perform an enterotomy if the object is in the intestinal tract. Exteriorize the segment of bowel containing the foreign body and isolate it by packing it off with moistened gauze sponges or laparotomy pads. Gently milk intestinal contents away from the site and have an assistant 'finger clamp' the bowel. Use sterilized 'bobby pins' (USA) or hair grips in the absence of an assistant. Make an incision in the antemesenteric portion of the bowel just distal to the foreign body. Remove the foreign body through this opening. Remove any necrotic bowel, if present, and anastomose healthy tissue. Close the bowel with 5–0 monofilament, non-absorbable suture if there is severe infection, otherwise a monofilament, absorbable synthetic suture material may be used. It is not advisable to use chromic catgut in this species. Lavage the bowel and replace it in the abdomen. Change sterile gloves and surgical instruments, and lavage the abdomen with warm sterile saline. Perform a routine closure of the abdomen with 4–0 or 5–0 monofilament, synthetic absorbable suture. Feed the ferret within 12–24 hours after surgery. Peritonitis is rarely a complication of a GI foreign body in ferrets. If peritonitis is observed, culture the abdomen for aerobic and anerobic bacteria and flush copiously. Treat with drain placement or leave as an open abdomen depending on the severity of the disease. Administer antibiotics based on culture results if peritonitis is present. Otherwise, place ferrets on a broad-spectrum antibiotic postsurgically.
c If the bowel is not necrotic and if severe peritonitis is not present, the prognosis is excellent.

103 A two-year-old female neutered ferret has been producing a green liquid mucus-containing stool for three days (103). The ferret has a decreased appetite, is moderately lethargic, 5% dehydrated and has a body temperature of 40.5°C. No abdominal pain is noted on palaption. Vomiting was seen three days ago, but has not been observed since. No other clinical signs are evident. A one-year-old neutered

male ferret adopted from a local ferret shelter was introduced into the household one week previously. The new ferret remains clinically normal. The ferrets are caged together, allowed only supervised exercise in the house and are fed a high-quality dry feline growth diet.

a What diseases would you consider for this ferret?
b What clinical investigations would you perform?
c How would you care for the ferret until a diagnosis is reached?

104 Two previously healthy pet rabbits housed together exhibit head tilt, ataxia, nystagmus and tremors within the same week in February. During mild weather they are allowed free run in the yard during the day and are brought indoors in the evening. Otoscopic examination was normal on both rabbits. The radiograph is of the skull of one of the rabbits (104).

a What are your differential diagnoses?
b Do the tympanic bullae appear normal in the radiograph?
c What other diagnostic tests or additional information from the history would help in establishing a diagnosis?

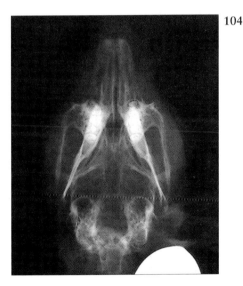

103 a Viral enteritis, bacterial enteritis, proliferative colitis, intestinal lymphoma, eosinophilic gastroenteritis, systemic disease such as renal or hepatic, intestinal parasitism and enteritis due to toxin ingestion.
b A fecal examination to rule out intestinal parasitism. Obtain hematology to assess immune response. Viral disease often results in a normal to depressed WBC count. Bacterial disease may cause a leukophilia with a neutrophilia. Lymphoma in the ferret can cause an elevation of the absolute lymphocyte count over the normal high of 3,500 or a 60% or higher lymphocyte portion of the differential. The peripheral absolute eosinophil count may exceed the normal high of 1,000 in ferrets with eosinophilic gastroenteritis.

Use radiography to assess intestinal gas patterns, liver size and any other abdominal abnormalities. Perform serum biochemistries to rule out other systemic disease. Fecal culture is occasionally helpful in diagnosing *Camplobacter* or *Salmonella* infections. In cases of suspected eosinophilic gastroenteritis, proliferative colitis or lymphoma, use endoscopy or exploratory surgery to obtain GI tissue samples for histopathology.

Conduct a thorough and detailed history to determine if there is any potential for toxin ingestion or exposure to other infected ferrets.
c If there is no vomiting, provide a combination of oral and subcutaneous electrolyte fluids to replace dehydration deficit and continue maintenance fluids (100 ml/kg q 24 hours). Often, ferrets will accept fruit-flavored oral electrolyte replacement solutions. Administer an oral broad-spectrum antibiotic, such as amoxicillin (15 mg/kg q 12 hours) or enrofloxacin (10 mg/kg q 12 hours), to cover the possibility of bacterial enteritis. Use a highly palatable, high calorie food supplement syringe fed in the amount of 10–20 ml/kg every 6–8 hours to maintain adequate calorie intake. In the USA, bismuth subsalicylate can be used as an intestinal protectant (17 mg/kg (1 ml/kg) q 12 hours). In the UK, alternatives are: BCK granules or Kaopectate (Kaogel: Parke Davis) (0.5 ml/kg q 12 hours). The primary source of calories should be fat as opposed to carbohydrates.

The diagnosis in the ferret in this case was probably viral enteritis contracted from the newly introduced ferret who experienced similar signs one month previously.

104 a Bacterial otitis media or interna, meningoencephalitis (bacterial, viral or proto-zoal), encephalomalacia (cerebral vascular accidents, parasites or toxins) and trauma.
b They are normal. In this case, normal tympanic bullae help to rule out chronic otitis media in which radiopaque caseous material and proliferative sclerotic bony change in the bullae are frequently seen. Tympanic bullae can appear normal in cases of acute otitis media.
c Hematology, serum biochemistries, cerebrospinal fluid tap and serology. Test results in this case are as follows: serum biochemistry values within normal limits, a WBC of 10,600 with a heterophil:lymphocyte percentage ratio (78:17) indicative of an inflammatory process. Cytology and culture of the cerebrospinal fluid were not diagnostic. Serology for *Encephalitozoan cuniculi* was negative. The outdoor history suggests a potential exposure to areas where raccoons might have been. Raccoons can pass *Baylisascaris procyonis* eggs in their feces and these eggs remain infective in the soil for years. When a non-host species, such as the rabbit, ingests the eggs, the larvae migrate to the brain where they continue to grow, causing progressive meningoencephalitis and encephalomalacia. Confirm the diagnosis of cerebral larva migrans by histopathology or isolation of larvae from brain tissue.

105

105 A breeding buck is in a rabbitry of about 100 New Zealand White rabbits (105). It has developed a mucopurulent nasal discharge and a slight head tilt.
a What is the cause of this condition?
b If the owner wants to save the buck, what treatment would you recommend?
c What recommendations would you make concerning the rest of the herd?

106

106 A wild red fox has developed generalized alopecia and marked thickening of the skin (106).
a What is your diagnosis for this skin condition?
b How would you treat it ?
c Is there a zoonotic potential with this disease?

107 An adult spider monkey has to be restrained for physical examination. The owner is loosely holding the monkey by a light-weight leash attached to a leather collar around its neck.
a What do you need to capture the animal?
b How should the owner assist you in its capture?
c How would you restrain a small- to medium-sized monkey?

105 a The most likely cause of upper respiratory disease and torticollis in this rabbit is infection with *Pasteurella multocida*.
b This disease can be controlled in many rabbits but it is difficult to achieve a complete cure and recurrence of clinical signs can occur when treatment is discontinued. Aggressive treatment early in the course of disease provides the best chance for eliminating a *P. multocida* infection in a patient. Chronically affected rabbits may require treatment for weeks to months. Obtain a deep nasal culture to aid in the appropriate antibiotic choice. Enrofloxacin (10 mg/kg PO or SC q 12 hours) or trimethoprim/sulfadiazine (30 mg/kg SC or PO q 12 hours) are good first-choice antibiotics with minimal side effects. Injectable procaine penicillin (40,000 iu/kg IM or SC q 24–72 hours for seven doses) can be used in rabbits that are established on a high fiber diet without current signs of GI disease. Follow penicillin injections with an oral antibiotic when long-term treatment is indicated. Do not use beta lactam antibiotics orally. Pasteurellosis can be precipitated by a number of factors including malnutrition, underlying disease, overcrowding, high environmental temperatures, high air ammonia levels and poor ventilation. Control of these variables will help to decrease the chance of disease.
c *P. multocida* infection is often endemic in a herd. Recommend that rabbits showing signs of disease be isolated and not bred. To establish a *Pasteurella*-free herd, test current and incoming breeders for *P. multocida* by culture and serology and breed only negative rabbits. Remove young from infected does by Cesarean section and foster them with *Pasteurella*-free does maintained under SPF conditions.

106 a Sarcoptic mange is very common in wild red foxes and is the most likely diagnosis in this case. Skin scrapings confirm the diagnosis of *Sarcoptes scabei*. Pruritus is often not evident. Other differential diagnoses to consider include dermatophytosis, bacterial dermatitis, other ectoparasite infestation and nutritional deficiencies.
b Ivermectin (0.2 mg/kg SC or PO) is the most effective treatment for sarcoptic mange in the fox. Repeat this dose at two-week intervals for one to two additional treatments. Keep the environment clean to prevent reinfestation. Do not release the patient back to its native environment until the treatment is completed.
c Sarcoptic mange has zoonotic potential for humans and a number of other animal species. Use scrupulous hygiene when handling wild animals with skin disease.

107 a The two most important items for capture are a net and leather gloves with gauntlets. The net should be light enough to avoid injury to the animal and have the appropriate mesh to prevent limbs from getting caught. The gloves need to be heavy enough to prevent bite wounds and they should extend to at least mid-forearm.
b Ask the owner to leave the room during the procedure. If the owner insists on staying, do not allow him/her to assist during the capture or restraint of the pet. Owners often think they can soothe the animal or that they can hold the monkey while you attempt to examine it. They cannot restrain frightened animals.
c Hold the monkey's arms behind its back with one hand and grasp the lower limbs for control with the other hand.

108

108 These two adult pet hamsters have developed chronic, non-pruritic generalized alopecia with scaling and crusts most evident around the ears and feet (108). The child in the household plays with the hamsters and the child was recently treated for dermatophytosis.
a What are your differential diagnoses for the hamster's skin changes?
b What diagnostic tests would you perform?
c How would you treat the hamsters?
d What recommendations would you make to the owner?

109a

109 Rabbits become sexually mature from three and a half to nine months of age. The smaller the breed, the earlier the onset of sexual maturity. Separate young rabbits into same sex groups by 10–12 weeks of age to avoid unwanted litters. The photograph shows five-week-old rabbits (109a).
a At what age can the kits be sexed and how would you do this?
b What is the gestation period of the rabbit?
c How soon can the doe be palpated for pregnancy?
d How should you cage a breeding pair of rabbits?

108, 109: Answers

108 a They include demodectic mange (*Demodex aurati, D. criceti*), dermato-phytosis, low protein (under 16%) feed, bedding abrasion, systemic disorders associated with ageing, including renal amyloidosis and renal neoplasia, staphylococcal pyoderma, adrenocortical neoplasia and other endocrine disease.
b Examine a deep skin scraping microscopically. Prepare a mineral oil or KOH wet-mount to assess for fungal hyphae. Culture the skin and hair for fungal and bacterial pathogens and evaluate the diet and husbandry practices. Perform further diagnostic tests as needed including radiography, ultrasonography, hematology, urinalysis and a serum biochemistry panel. A heavy infestation of *Demodex* spp. was identified in this case. Additionally, abundant growth of *Trichophyton mentagrophytes* on dermatophyte test medium was observed.
c Attempt to treat demodectic mange with ivermectin (0.2–1.0 mg/kg SC). Repeat the ivermectin twice at 10–14 day intervals. This treatment often does not work because ivermectin may not be effective. Use of amitraz is dangerous in hamsters because even when diluted it is extremely toxic to this species. Usually *Demodex* spp. become clinically apparent when there is underlying primary disease such as adrenal gland disease, neoplasia, severe environmental stress, malnutrition and other systemic disease. Demodocosis is not zoonotic. *T. mentagraphytes* or *Microsporum canis* are the most common causes of dermatophyte infections in hamsters. Dermatophyte infections on hamsters can be zoonotic. Treat dermatophytosis with griseofulvin (25–30 mg/kg PO q 24 hours for 3–4 weeks). Treat localized infections with a topical antifungal cream or lotion.
d Completely clean the cage and use non-toxic pelleted bedding. When handling, wear gloves or wash hands with a germicidal hand soap afterwards.

109b

109 a The kits are best sexed at birth or at weaning (five to eight weeks of age). In the interim, it is difficult to exteriorize the genitalia. Stretch the perineum to expose the anogenital structures. The male has a conical- to cylindrical-shaped penis with a rounded or oval urethral opening. The female has a vulva that protrudes slightly with a slit-like opening (109b). In sexually mature males, the scrotal sacs may be seen lateral to the perineum. Male rabbits can retract the testicles into the inguinal canal making sex determination more difficult.
b Twenty-nine to 35 days, with a litter size ranging from 4 to 12.
c Fetuses can be palpated as early as 10 days postbreeding. At this stage they can be felt as masses approximately 1–1.5 cm in diameter located in the caudal ventral abdomen. At 18 days postbreeding, the fetuses are approximately 2.5–3 cm in length. Palpation must be gentle to avoid damage to the fetuses. If necessary, use radiography or ultrasonography after 21 days to determine pregnancy.
d Does are territorial and may kill a new rabbit introduced into the cage. For this reason, it is important to bring the doe to the buck's cage for breeding. If the doe does not accept the buck within a few minutes and fighting occurs, separate them. Introduce them again at 12–24 hour intervals until one to two successful matings have occured.

110 A dental abnormality is seen as an incidental finding during a preoperative examination on a pet rat prior to excision of a mammary tumor. A lateral radiographic view of the patient's skull is shown (110).

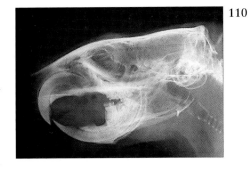

a How does the anisognathism of rodents differ from that in carnivores, rabbits, horses and cattle?
b What is the dental formula for rats and mice?
c How does their dentition differ from that of guinea pigs and chinchillas?
d What dental abnormality is shown in the radiograph and what are the three most likely causes?

111 A three-month-old raccoon kept by a wildlife rehabilitator for four days has bloody, foul-smelling diarrhea of 24 hour duration. The animal is severely dehydrated and depressed. Hematology reveals a marked anemia and leukopenia. Ascarid ova are found on a fecal examination. The rehabilitator has 14 other raccoons at her home that have been kept isolated from the new raccoon.
a What are the differential diagnoses?
b What is the most likely diagnosis?
c How would you manage the raccoon?

112 A mouse has a three-week history of scaling and pruritus (112). Skin scrapings and acetate strip samples are negative for ectoparasites.
a What is your diagnosis?
b How would you confirm this diagnosis?
c Is there a zoonotic risk?

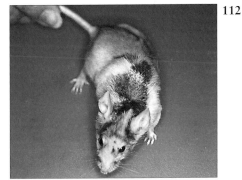

110 a In rodents the maxilla is narrower than the mandible, providing space for the large masseter muscles which are required for gnawing. In carnivores and herbivores, such as rabbits, horses and cattle, the jaw width relationship is reversed with the mandible being narrower than the maxilla.

b $I\underline{1}$ $C\underline{0}$ $P\underline{0}$ $M\underline{3}$
      1   0   0   3

c Rats and mice are granivorous rather than herbivorous like chinchillas and guinea pigs. The granivore diet is nutritionally concentrated, is relatively non-abrasive and requires little grinding before it is swallowed. The cheek teeth are reduced in number and size compared to the herbivorous species. Rats and mice have brachyodont molar teeth that stop growing once they have erupted and the roots are fully formed.

d One of the rat's mandibular second molar teeth is missing. Absence of molar teeth may be congenital, the result of periodontal disease or post-extraction. Caries is not likely to be the cause because it is rare in rodents unless they are fed an unnatural diet and are exposed to cariogenic bacteria. Loss of molars from external trauma is also unlikely because they are protected by a thick layer of masseter muscle laterally.

111 a Canine distemper, raccoon parvovirus, parasitic enteritis, bacterial enteritis and diarrhea due to dietary changes.

b Raccoon parvovirus. Raccoon parvovirus is related to canine parvovirus, mink enteritis virus and feline panleukopenia. Raccoons are not susceptible to canine parvovirus and raccoon parvovirus is species-specific. Clinical signs of raccoon parvovirus include depression, anorexia, dehydration, diarrhea and leukopenia. The diarrhea may be watery, mucoid or hemorrhagic. It is the most common cause of severe diarrhea associated with leukopenia in wild raccoons. Confirm the diagnosis by submitting blood or feces for IFA or ELISA testing.

c This disease is highly contagious to other raccoons and has a 80–100% mortality rate. The incubation period is approximately seven days with death occuring 10–14 days after clinical signs are noted. Isolation and euthanasia of infected animals is usually indicated. Follow the same treatment protocol used for dogs with parvovirus infections. Use strict hygiene when handling the fecal material of wild raccoons due to potential presence of the larvae of *Baylisascaris procyonis*. The ascarid ova found on the fecal examination of this patient are most likely those of *Baylisascaris*. This parasite does not usually cause clinical signs in the host animal; however, when the ova are ingested by non-host species, including humans, the larvae that emerge can cause severe damage and even death by visceral larva migrans.

112 a Dermatophytosis is the most likely diagnosis. The distribution of lesions is not pathognomonic, although facial lesions suggest dermatophytosis. *Trichophyton mentagrophytes* is frequently isolated in cases of murine dermatophytosis.

b Mount affected hairs in mineral oil and examine them microscopically. A KOH preparation can also be used, but some hairs will appear positive if they are left in this preparation for an extended period of time. *Trichophyton* spp. do not fluoresce, therefore a Wood's lamp examination is of no value. Confirm the diagnosis by submitting hairs from the affected areas for fungal culture.

c Yes. The risk to humans is greater following exposure to infected reservoir host species, such as pet mice and wild mice, than from exposure to an infected dog or cat. However, the likelihood of human infection is less than that from a *Microsporum canis* infection in a cat.

113 A six-month-old pet intact female rabbit is exhibiting increased frequency of urination. She eliminates small amounts of urine in several selected areas of the room where she is housed. The urine is reddish in color. The tables below show the results of urinalysis and serum biochemistries.
a What are your differential diagnoses for 'red urine'?
b What is your diagnosis in this case?
c How would you treat the various causes of 'red urine'?

Urinalysis results

Appearance	Cloudy
pH	8.5
Specific gravity	1.02
Protein	Trace
Glucose	Negative
Ketone	Negative
Bilirubin	Negative
Blood	Negative
WBC	Rare
RBC	None
Casts	None
Epith.	Rare
Crystals	Calcium carbonate and triple phosphate
Bacteria	Rare

Serum chemistry results

Glucose	7.56 mmol/l (135 mg/dl)
BUN	18 mg/dl
Creatinine	106.08 µmol/l (1.2 mg/dl)
Sodium	140 mmol/l (140 mEq/dl)
Potassium	4.5 mmol/l (4.5 mEq/dl)
Calcium	3.55 mmol/l (14.2 mg/dl)
Phosphorus	1.152 mmol/l (3.6 mg/dl)
Total protein	57 g/l (5.7 g/dl)
Albumin	37 g/l (3.7 g/dl)
Globulin	20 g/l (2.0 g/dl)
Bilirubin total	4.5 g/l (0.45 g/dl)
Alk. phos.	46.5 U/l
SGPT/ALT	37.3 U/l
SGOT/AST	18.7 U/l
GGT	5.5 U/l
Chloride	108 mmol/l (108 mEq/l)
CO_2	20.7 mmol/l (20.7 mEq/l)

114 A squirrel monkey with a five-day history of watery diarrhea is now 8–10% dehydrated. A fecal floatation using sodium nitrate reveals this parasite identified by trichrome stain (114).
a What is the organism?
b Could this be the causal agent of the diarrhea?
c How would you treat the infection?
d What are the difficulties associated with administration of the drug of choice?

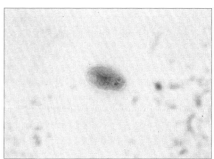

114

115 An eight-month-old female hamster has non-specific signs of lethargy and ataxia of 24–36 hour duration. Since hamsters hoard food, the owner is not sure about her appetite, but she appears of normal body condition and approximately 3–5% dehydrated.
a What is the proper procedure for capture and restraint of the hamster?
b What are your differential diagnoses?

113 a It can be the result of blood from the urinary or female reproductive tract, dietary pigments or porphyrin pigments. Hematuria occurs with urinary tract infection, trauma, neoplasia, urolithiasis, bladder polyps or reproductive disease, such as uterine aneurysm, adenocarcinoma or pyometra. Blood from the uterus pools in the vagina and is expelled when the rabbit urinates. Uterine bleeding from adenocarcinoma is sporadic and often appears as clots of blood within the voided urine. Rabbits can normally eliminate porphyrin pigments in the urine. These pigments can color the urine dark yellow to a deep red-orange. The factors that cause porphyrin production are unknown. Some foods can cause urine color changes. Most notable are beets which create urine that has a magenta hue.
b Both the serum biochemistry and the urinalysis values are normal. The red color of the urine is due to porphyrin pigments. The frequency of urination in this sexually mature animal is consistent with territorial marking.
c Treat cystitis and pyelonephritis with aggressive antibiotic and fluid therapy. Remove uroliths and polyps surgically. Perform an ovariohysterectomy for uterine disease. Adenocarcinoma is the most common uterine disease in the adult female domestic rabbit. Radiograph or ultrasound the patient prior to an ovariohysterectomy to screen for metastatic disease. No treatment is necessary for urine colored by dietary or porphyrin pigments. An ovariohysterectomy was performed on this case to curtail the marking behavior. The frequent urination stopped approximately three weeks post-surgically. In some cases of territorial marking, additional behavior modification may be necessary.

114 a *Giardia lamblia.*
b Yes. Since other factors may be involved in the diarrhea, it is important to include a CBC and serum biochemistry panel on this monkey. Other tests include examination of fecal smears and fecal floats and fecal cultures.
c Metronidazole (15–30 mg/kg/day PO q 24 hours for 5 days). The disease is usually self-limiting but treatment shortens its course. Warn the owners that this disease has zoonotic potential.
d Only the oral form of metronidazole is effective and it has an extremely bitter taste. It is not possible to 'pill' a primate. Many primates can detect the taste of metronidazole, even in small quantities, in any type of food. Manual restraint is often necessary to administer the drug via nasogastric tube.

115 a If the hamster is relatively tame or debilitated, gently scoop it up in a cupped hand. To avoid being bitten by a fractious patient, scoop it into a can or small box. Do not pick up hamsters when they are sleeping, since many will bite in response to being surprised. Once the hamster is captured, carefully scruff it by grasping the large skin fold over the nape of the neck. This skin is loose and plentiful, and if a sufficient amount is not grasped, the pet can turn and bite. Reposition your hold as needed to minimize pressure that may cause the eyes to bulge.
b Infectious disease, metabolic conditions (particularly renal or hepatic disorders), neoplastic disease, digestive disorders (dental malocclusion, gastric foreign bodies), trauma and toxicity. The described signs are non-specific, therefore further investigation is necessary.

116 How would you treat severe, life-threatening anaphylactic vaccination reactions in ferrets?

117 A two-year-old ferret jill is dull, lethargic and has a generalized alopecia (**117**).
a What are your differential diagnoses for the alopecia?
b How would you treat these conditions?

118 The combination of ketamine and medetomidine provides good surgical anesthesia in several species of small mammals (**118**). What other particularly important advantage does it have?

116 Both canine distemper and rabies virus vaccines have been implicated. Most anaphylactic reactions occur within 15 minutes of vaccine administration. A variety of signs can be seen including clear to blood-tinged vomit, bloody diarrhea, dyspnea, cyanosis, tachycardia and collapse. If the patient is not immediately treated, seizures, coma and finally death can occur. Administer oxygen by endotracheal tube or facemask and place an intravenous catheter. Administer epinephrine (0.25–1.0 ml of a 1:10,000 dilution IV) and diphenhydramine hydrochloride (1.0–2.0 mg/kg IV). In addition, give either hydrocortisone sodium succinate (25–40 mg/kg IV), or prednisolone sodium succinate (5 mg/kg IV) along with lactated Ringer's solution with added dextrose. If the intravenous route cannot be established, give the above medications intraosseous or intramuscular. Advise the owner to remain with their animal at the veterinary hospital for at least 30 minutes following vaccination to observe for anaphylactic reactions. Animals that have had previous vaccine reactions should either not receive future vaccinations or be premedicated with diphenhydramine hydrochloride at least one hour before vaccination and observed for at least one hour afterwards.

117 a Adrenocortical disease (hyperplasia or neoplasia) and hyperestrogenism are the most likely differential diagnoses. In the ferret, disease of the adrenal gland is not related to the pituitary as in other species. It is usually a disease of ferrets three years of age and older, although it has been reported in animals as young as 18 months. Ferrets affected with adrenal disease may be either intact or neutered. The alopecia in adrenocortical disease in the ferret is caused by an increased production of androgens rather than an increase in corticosteroids as in other animals. Some neutered female ferrets will exhibit an enlarged vulva as if in estrus in response to the abnormal androgen production. Hyperestrogenism occurs when an intact jill remains unbred and in estrus for an extended period of time. The most prominent sign of estrus is vulvar swelling. The effects of prolonged exposure to high levels of endogenous estrogen are alopecia and bone marrow suppression. Signs include petechial hemorrhages on the skin and mucous membranes, pallor, lethargy and generalized weakness.
b Manage adrenocortical neoplasia with adrenalectomy of the affected gland(s) or mitotane therapy (see question **212** for a discussion of treatment). Obtain a CBC and platelet count on ferrets with hyperestrogenism to determine the extent of disease. If the PCV is 15% or lower, the prognosis for recovery is grave. Terminate the estrus cycle with hormonal therapy (HCG at 100 iu, IM or GnRH at 20 mcg/ferret SC), mate with a vasectomized male or an ovariohysterectomy. Hormonal therapy may not be effective in the moderate to severely affected case, therefore perform an ovariohysterectomy as soon as the patient is stable. Ferrets have no discernable blood types, therefore it is possible to use multiple donors when performing blood transfusions. Provide other supportive care as needed. Attempt to stimulate the bone marrow with anabolic steroids or synthetic erythropoietin used at feline dosages.

118 The medetomidine component of the anesthetic mixture is rapidly and completely reversed using atipamezole. Since ketamine, when used alone, has relatively little sedative effect in small rodents, reversal of medetomidine produces a rapid lightening of the plane of anesthesia and often a rapid return to consciousness. This is a particular advantage in small mammals since it reduces the likelihood of hypothermia developing in the postoperative period.

119 A pet hamster has fallen off a table and fractured its incisor teeth (**119**). The pulp cavity is not visibly exposed.
a What is the dental formula of the Golden hamster?
b The hamster has typical rodent incisor dentition, but what form do the cheek teeth take?
c How would you treat the fractured right lower incisor?

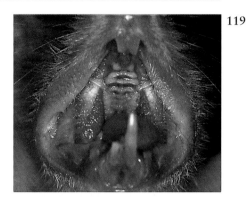
119

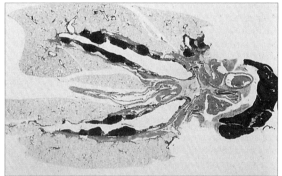
120a

120b

120 A two-year-old pet female rat exhibits wheezing respiratory sounds and dyspnea. The rat is euthanized and the lungs examined (**120a, b**).
a Describe the gross pathology and histopathology.
b What is the likely etiology?
c What treatments can you use for this disease and what is the prognosis?

119, 120: Answers

119 a   I $\underline{1}$ C $\underline{0}$ P $\underline{0}$ M $\underline{3}$

            1   0   0   3

Some sources classify the first cheek teeth as premolars not molars.
b Hamsters have brachyodont cheek teeth similar to rats and mice. Also, like rats and mice, hamsters are granivorous rather than herbivorous.
c Administer analgesics immediately post-trauma. No other immediate treatment is needed if the pulp chamber has not been exposed. Shorten the right upper incisor at regular intervals (avoiding iatrogenic pulp exposure) until the lower tooth has grown back into normal occlusion. Unfortunately, in some cases, the trauma damages the germinal tissues at the root apex of the damaged tooth and it does not regrow. In these cases continue to trim the opposing tooth or extract it. Use a dental burr in a high-speed dental hand piece to trim the incisors. Do not trim rodent or rabbit teeth with nail cutters. This practice commonly leads to iatrogenic pulp exposure which can ultimately result in more serious tooth root disease.

120 a Grossly, the lungs show bronchopneumonia and areas of atelectasis due to bronchial occlusion. The pleural surface is unevenly elevated due to bronchiolectasis. Histopathologically, the lung tissue reveals peribronchial lymphoid hyperplasia and neutrophil exudation within the bronchial lumina.
b These lesions are typical of murine respiratory mycoplasmosis caused by *Mycoplasma pulmonis*. Infection is transmitted both horizontally and vertically and is chronic for the life of the animal. It is enzootic in many rat colonies unless they have originated from hysterectomy derivation. The disease may lie dormant for varying periods and asymptomatic carriers exist. The agent may be isolated from tracheal washings; however, special enriched media is required and subculturing is often necessary. Most laboratories are not equipped to isolate *M. pulmonis*. A positive antibody titer to *M. pulmonis* is indicative of infection.
c Treat rats with suspected *M. pulmonis* infections with one or a combination of these antibiotics: enrofloxacin (15 mg/kg PO q 12 hours), chloramphenicol (50 mg/kg SC q 8 hours or 1 g/l drinking water), tetracycline (20 mg/kg PO q 12 hours), oxytetracycline (60 mg/kg SC q 72 hours) or tylosin (10–20 mg/kg IM q 24 hours or 100 mg/kg PO q 8–12 hours). It is difficult to achieve a complete cure and relapse is common. The prognosis for controlling mild disease in a young rat is good. The success of treatment decreases as the age of the rat, the duration of the disease and the likelihood of secondary infection increase. Radiography will aid in determining the extent of pulmonary involvement and prognosis.

121 Ovariohysterectomy is a common procedure in the pet rabbit.
a Why would you recommend ovariohysterectomy in non-breeding rabbits?
b What are some special considerations of ovariohysterectomy in a rabbit?
c How would you perform an ovariohysterectomy in a rabbit?

122a

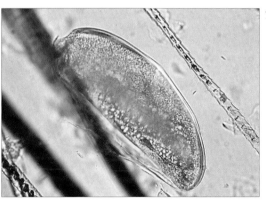

122b

122 This mouse (**122a**) has had pruritus for four weeks. Acetate tape strippings reveal only these objects (**122b**).
a What are the objects?
b What produced them?
c How would you treat the condition?

121, 122: Answers

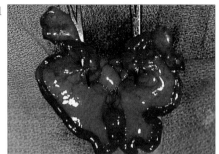

121 a Uterine adenocarcinoma is found in 30–80% of intact does over three years of age, so ovariohysterectomy is a preventive health measure. Ovariohysterectomy may prevent unwanted sexually related behavior, such as territoriality and urine marking.

b Prevent bruising and postoperative complications by observing careful handling of the delicate rabbit skin and tissue. The mesovarium, mesosalpinx and mesometrium are often very fatty and friable in mature rabbits and tear easily if too much tension is applied. Do not use clamps on these tissues. The rabbit has a duplex uterus with two uterine horns and no uterine body (**121**). The vagina extends cranially, ending at the double cervix. Identify the vagina properly and do not mistake it for a (non-existent) uterine body.

c In young rabbits make a small (2–4 cm) ventral midline incision centered between the umbilicus and the rim of the pelvis. Create a larger incision if needed in older animals with more developed reproductive organs or uterine pathology to fully exteriorize the ovaries and uterus. Empty the bladder prior to surgery and avoid the voluminous bowel when making the abdominal incision. Identify the duplex uterus dorsal to the bladder in the caudal abdomen and elevate it without the use of a spay hook. Isolate one ovary and double ligate the branching ovarian vessels with synthetic absorbable suture. Identify and include the associated uterine tube and infundibulum with the tissues to be removed. The mesometrium contains many vessels distributed in a fan-like pattern. These are difficult to identify in fat animals, but tie them off by ligatures placed in the mesometrium. Identify and double ligate large uterine vessels close to the cervix on each side. Clamp, ligate and remove each uterine horn as close to the cervix as possible. Alternatively, clamp and ligate the vagina just caudal to the double cervix to assure that all uterine tissue is removed. Oversew the vaginal stump to prevent the leakage of urine into the abdomen. Close the abdominal wall with synthetic absorbable suture. Close the skin with synthetic absorbable suture in a buried continuous intradermal pattern or with skin staples.

122 a The eggs of parasitic mites.

b They come from surface-living mites, such as *Myobia musculi*, or pelage mites such as *Myocoptes musculinus*. Lesions are usually more severe with the surface living parasites, probably due to the increased possibility of hypersensitivity reactions.

c Treat surface-living mites with ivermectin (0.2–0.4 mg/kg PO, SC or topically). Repeat the treatment once or twice at two-week intervals. Use topical insecticides for pelage mite infestations because ivermectin is ineffective. Repeat these treatments with the same frequency as described for surface mites. Exercise great caution when using topical insecticides because of potential absorption of toxic amounts through the mouse's thin skin or through grooming. Avoid using these products in debilitated animals. Treat all mice within the same cage. Thoroughly clean the environment at frequent intervals during treatment.

123 An albino adult female mouse was anesthetized with a combination of ketamine and xylazine to facilitate removal of a skin tumor. Approximately 30 minutes after anesthesia induction, both eyes began to appear white. After one hour, the opacity progressed to involve most of the pupillary region of the eye (**123**).
a What ocular changes are apparent?
b What caused these changes?
c What is the prognosis?

124 A two-year-old neutered male rabbit has these intermittent thick, pudding-like stools during the day (**124**). The soft stools are produced daily along with normal-sized dry droppings. The rabbit is fed a commercial rabbit pellet (16% fiber, 14% protein) free choice and is given a teaspoon a day of rolled oats and one salt free cracker.
a Why is this diet a likely cause for the soft stools?
b What is the preferred diet for the adult non-breeding pet rabbit?

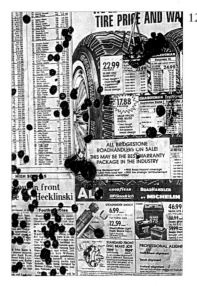

123 a There is an axial cataract primarily of the anterior cortical region of the lens.
b Xylazine is implicated as the causative agent of cataracts, although the exact mechanism by which xylazine produces cataracts in rodents is unclear. Typically, the cataracts begin as fine white opacities around the anterior 'Y' sutures of the lens, often followed by a diffuse, gray-white opacity that may spread to involve the entire lens. Topical application of xylazine also produces similar lens opacities. Closing the eyelids to prevent ocular surface drying appears to prevent cataracts from forming. Therefore, it has been suggested that the opacities result from tear-film irregularities and/or ocular surface drying which leads to relative dehydration of the anterior chamber, disorganization of lens proteins and subsequent cataract formation. This explanation is plausible due to the relatively large lens and very shallow anterior chamber in rodents. Evidence for this hypothesis is found in the photograph because the normally circular ring flash artefact from the central cornea is grossly distorted suggesting the ocular surface has dried. Alternatively, xylazine induces cataracts in rodents by directly affecting the hydration of the lens epithelial cells. It can do this by either interfering with secretory cells that make components of the tear-film or by impairing the function of the cells that regulate water movement across the cornea. Finally, alpha-2 adrenergic agonists (i.e. xylazine) are potentially potent ocular hypotensive agents and it is possible that the cataract results from a reduction in aqueous humor production and subsequent relative dehydration of the lens proteins.
c The opacity usually persists from one to two hours and then usually completely resolves within several hours to a day.

124 a Low-fiber diets (<18%) create cecal–colonic hypomotility. This predisposes the animal to abnormal cecal fermentation due to prolonged retention of digesta in the cecum. Low-fiber diets and those high in carbohydrates stimulate production of volatile fatty acids resulting in pH alterations, ultimately changing the cecal microflora. This loss of intestinal homeostasis causes soft stools in the rabbit. The soft stools are actually liquid cecal contents which should have formed cecotropes that the rabbit would normally ingest. Other signs include obesity and GI ileus (sometimes incorrectly identified as a 'gastric hairball'). Simple sugars and starches provide an environment in which pathogens such as *Escherichia coli* and *Clostridium* spp. proliferate. The addition of dietary glucose causes the production of iota toxin by *C. spiroforme*, the bacteria responsible for enterotoxemia in the rabbit. Diets high in fiber seem to have a protective effect against enteritis. The beneficial effect is associated with the indigestible fiber component, lignocellulose. Digestible fiber sources do not have this property. Fiber stimulates cecal-colonic motility. Other effects of fiber are indirect. High-fiber diets naturally have a lower carbohydrate concentration, which decreases the chance of caborhydrate overload enterotoxemia.
b Offer adult pet rabbits a good quality grass hay (marsh, orchard, timothy or Bermuda) *ad libitum*. The protein and calcium content of alfalfa is greater than that required by the adult rabbit. Feed a good-quality, high-fiber pellet (18–24% fiber) in limited amounts to maintain normal body weight. Most adult non-breeding rabbits can be maintained on unlimited grass hay and fresh herbage without pellets in their diet. Offer fresh herbage at a rate of a minimum of 1 cup/kg body weight daily. Do not feed foods high in starch (such as grain) or refined sugar.

125 What immediate treatment would you give a rodent or rabbit for fractured incisor teeth with pulp exposure?

126 A diagram of the rabbit GI tract (**126**). Name the labelled structures and describe the special or unique features of each.

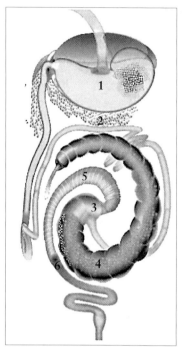

126

127 A client buys two hamsters from a pet store. Three weeks later she discovers a litter of nine newborns in the cage. She has several questions that she would like you to answer.

a Why did the pet store owner not tell her that one of the hamsters was pregnant when she purchased them?

b She was told that she had two female hamsters. How can she tell what sex they are?

c Should she leave the two original hamsters together to share in the rearing of the young ?

125 Pulp exposure, whether traumatic or iatrogenic in origin, requires proper treatment to relieve pain and minimize later complications, such as chronic pulpitis or root abscessation. Clean and disinfect the fractured tooth with dilute chlorhexidine or a similar agent. In the smallest species (mice, gerbils and hamsters), dry the fractured tooth surface and apply a setting calcium hydroxide cement to seal the pulp exposure. The very rapid metabolic rate of these species means that secondary dentine is quickly laid down to seal the exposure site providing the pulp remains viable. Administer antibiotics and analgesics as necessary. In larger species, perform a partial pulpectomy. Make a slight undercut in the dentine, using a small round burr and aseptic technique. Once pulpal hemorrhage has stopped, apply a small amount of calcium hydroxide powder directly on to the surface of the remaining pulp. Use setting calcium hydroxide cement to restore the tooth surface. This cement is usually strong enough to remain in place while healing takes place. In rabbits, make a deeper preparation and place a stronger restoration over a layer of calcium hydroxide cement. Take care in selecting a restorative material in this situation as it must not interfere with normal tooth wear or release toxic substances as the tooth and restoration wear away. Recontour the fractured tooth surface to remove sharp edges to make it more functional.

126 1 Cardia of the stomach. This area has a well developed sphincter that prevents vomiting. The fundus is exocrine and secretes acid, intrinsic factor and pepsinogen. The pH of the stomach of adult rabbits is usually <2.
2 Pancreas. The pancreas is diffuse in the mesentery of the small intestine. Multiple ducts empty into the terminal duodenum separately from the bile duct.
3 Sacculus rotundus or cecal tonsil. This is the terminal part of the ileum and contains many lymph follicles.
4 Cecum. The cecum comprises 40% of the GI tract of the rabbit. It is the organ where the major bacterial fermentation of cellulose occurs.
5 Proximal colon. This is the first limb of the colon having rows of sacculations or haustrae which control movement of ingesta out of or back into the cecum.
6 Fusus coli. This is a heavily innervated structure that regulates the passage of material into the descending colon. Retrograde peristalsis transports ingesta with less fiber back to the cecum where cecotropes, containing vitamins, minerals and amino acids are formed. Cecotropes are eliminated once or twice a day and ingested directly from the anus.

127 a The gestation period in hamsters is 16 days, therefore the hamster was not pregnant when purchased. Apparently, the owner has a hamster of each sex that have successfully mated since they were acquired.
b It can be difficult to accurately determine the sex of young hamsters. The most reliable way to sex them is by comparing the anogenital distance. Males have a longer anogenital distance than females. In addition, male hamsters over three months of age have prominent scrotal sacs that extend caudally as they mature.
c Male hamsters often cannibilize the young and female hamsters may aggressively attack the male to protect her young. Postpartum estrus is usually infertile, but the female will ovulate soon after the young are killed or removed. Recommend separating the pair to prevent injuries and rebreeding. Remove the babies at three weeks of age from their mother to prevent injury or death.

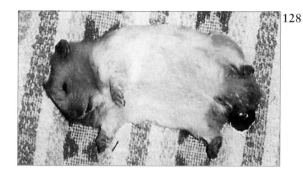

128 A hamster is sedated for examination of a perianal mass (128). The mass is an abscess from which *Staphylococcus aureus* is isolated in pure culture.
a What factors should you consider when deciding on antibiotic treatment of this hamster?
b How would you manage this case?

129 A lateral radiographic view of the caudal skull of a common marmoset (129).
a What metabolic bone disease is demonstrated in this radiograph?
b What clinical signs might you see when this condition is present?
c How would you confirm the diagnosis?
d How would you treat the condition?

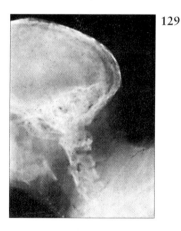

128 a Exercise great caution when using antibiotics in hamsters. They are very sensitive to antibiotic-induced enterocolitis, which can be fatal. Mortality has been reported after oral or parenteral use of many antibiotics including: penicillin, ampicillin, cephalexin, cefoxitin, cephalothin, carbenicillin, lincomycin, clindamycin, streptomycin, gentamicin, neomycin, chloramphenicol, erythromycin, tetracycline, vancomycin and some trimethoprim-sulfamethoxazole preparations. Other species where antibiotic-induced enterocolitis is a significant problem include the rabbit, guinea pig and chinchilla. Rats, mice and gerbils are rarely affected. Use enrofloxacin in species where antibiotic-induced enterocolitis is of concern. Studies have shown it to be safe in guinea pigs and rabbits, and anecdotal evidence suggests it is safe in the hamster. The dose rate in the larger species is 5–10 mg/kg with ranges up to 85 mg/kg suggested in smaller species, such as mice. Consider the possibility of cartilage damage if enrofloxacin is used in growing animals.
b Surgically debride the abscess, flush with an antiseptic solution and apply an appropriate volume of a topical non-absorbable antibiotic. Mupirocin would be suitable in this case, because it is not absorbed and is active against staphylococci. Use an effective systemic antibiotic as indicated by the bacterial sensitivity and one that is safe, such as enrofloxacin or trimethoprim/sulfamethizole.

129 a Bone resorption caused by excessive secretion of parathyroid hormone due to secondary nutritional hyperparathyroidism, which is often seen in this species. In this case, there was insufficient dietary intake of vitamin D_3 (New World primates require vitamin D_3, not D_2) which produced insufficient intestinal absorption of calcium. As blood calcium concentration falls, parathyroid hormone concentration rises to mobilize calcium from the bones into the circulation. Although low calcium intake and kidney disease can also cause this type of bone resorption, the principal cause in captive common marmosets is lack of dietary vitamin D_3.
b When bone resorption becomes severe, captive primates become inactive and move slowly about their cages due to pain, muscle weakness and pathologic fractures. The teeth become loose resulting in eating difficulties.
c Radiograph the caudal skull and long bones. If possible, obtain a normal long bone or skull taken from a healthy monkey at necropsy. Take a radiograph with the normal bone placed next to the patient. A comparison of bone density can be made that is independent of between-film variations in exposure. Obtain a blood sample to detect low serum calcium concentration. An ionized calcium concentration is a more sensitive test than total serum calcium. Absolute confirmation of the diagnosis is possible by measuring blood parathyroid concentration. Some test kits used in humans are suitable. Parathyroid hormone concentration up to 10 times normal can be seen.
d Although the need for vitamin D_3 rather than vitamin D_2 is commonly understood, the high dietary requirement in this species is less recognized. Marmosets and some other New World primates have a 1,25 (OH)2 vitamin D_3 receptor that is either blocked by a protein or is much less sensitive to vitamin D_3. The blood concentration of vitamin D_3 needs to be high to compensate for the presence of this blocking protein. Maintain a high blood concentration of vitamin D_3 in the marmoset by administering approximately 1,100 iu/kg/day (compared to 11 iu/kg/day in dogs) of vitamin D_3. In addition, expose New World primates to unfiltered sunlight to encourage production of vitamin D_3 in the skin. It is may be possible to provide some UV light using daylight spectrum light bulbs.

130 An adult rat has been exhibiting signs of respiratory disease for several weeks. It was treated unsuccessfully with oxytetracycline in the drinking water for one week. The owner elects euthanasia due to the rat's extreme respiratory distress. Multiple abscesses are present in the lungs of the rat at gross necropsy (130).

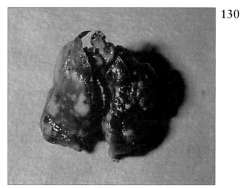

a What agents are responsible for these lung lesions?
b Was antimicrobial therapy a rational choice in this case?
c If antimicrobial therapy was indicated, was oxytetracycline a suitable agent?
d Is administration of oxytetracycline in the drinking water an effective route of administration?

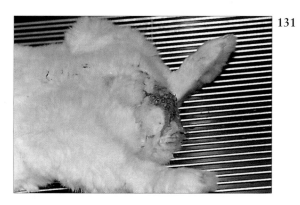

131 A five-year-old intact adult female rabbit has a decrease in appetite and exhibits perineal staining (131). She is slightly overweight and the owner notices that she makes a sound as if she is grinding her teeth periodically throughout the day.
a What are the differential diagnoses for the perineal staining?
b What diagnostic procedures should you perform?
c Why should non-breeding does be spayed?
d At what age should you perform the procedure?

130 a Multiple abscesses are commonly seen as the end stage of several infectious diseases. The likely etiologic agents are *Corynebacterium kutscheri*, *Mycoplasma pulmonis* and the cilia associated respiratory (CAR) bacterium. Although each of these agents is individually capable of causing disease, it is likely that more than one is present. There may also be an underlying infection with Sendai virus or pneumonia virus of mice.
b Antibiotic therapy is a rational choice in individual rats kept as pets because of the complex possible etiology with a high likelihood of bacterial involvement.
c Oxytetracycline is considered one possible therapy for mycoplasmal infections, and is often effective against corynebacteria. Antibiotic susceptibility of CAR bacillus has not been investigated. Oxytetracycline was a rational therapeutic choice in this case.
d There is a common misconception that oral administration of tetracyclines is effective for systemic therapy of rodents and rabbits. It has been conclusively shown in rats, and also in rabbits, that oxytetracycline is not absorbed after oral administration. It is likely that absorption is also inadequate in other rodents. One reason oxytetracycline is administered in the drinking water is to ensure compliance for long-term treatment. A more effective route of administration is to use injectable, long-acting preparations of oxytetracycline (60 mg/kg q 72 hours). Use subcutaneous injections due to the large volume of drug needed. There is no advantage to intramuscular administration over subcutaneous administration in terms of absorption. Different preparations of long-acting oxytetracycline have been shown to vary greatly in the degree of tissue irritation and blood concentration they induce.

Consider changing antibiotics if there is no response to treatment within a few days. Other choices include oral or parenteral tylosin, chloramphenicol tetracycline or enrofloxacin. Respiratory disease in rats can be difficult to eradicate completely due to multiple etiologic agents, resistance to antibiotics and carriers in the colony. Reoccurrence of signs is common after antibiotics are discontinued.

131 a They include uterine disease (pyometra, neoplasia, dead fetuses), pregnancy, urinary disease (urolithiasis, neoplasia, cystitis, incontinence), neurologic disease or obesity preventing grooming and a dirty cage environment. Excessive tooth grinding often accompanies conditions that are painful, such as dental or abdominal disease.
b Radiograph the rabbit to assess the reproductive and urinary tracts and to detect pulmonary metastasis of neoplastic disease. Continue with other diagnostic testing as needed including ultrasonography, serum biochemistries, hematology, urinalysis and urine culture. Use exploratory surgery as a diagnostic tool or as a treatment. This patient was diagnosed with uterine adenocarcinoma without apparent pulmonary metastasis. The vaginal discharge and tooth grinding resolved after an ovariohysterectomy was performed.
c Because a large percentage of rabbits develop some form of uterine disease by the time they are middle-aged. The most common uterine disorder is adenocarcinoma. This malignant neoplasia has the potential to metastasize. Other uterine diseases include pyometra, mucometra, endometritis and uterine aneurysm.
d Spay rabbits between four months and two years of age to avoid uterine disease.

132 European hedgehogs are frequently victims of road traffic accidents and are commonly presented for treatment. If such casualties are to be released back into the wild, it is important to ensure that they will be able to survive.
a What is the adult dental formula for this species?
b What treatment would you give for fractured teeth?
c How would you stabilize a bilateral mandibular fracture in normal occlusion without adding to the accompanying soft-tissue trauma?

 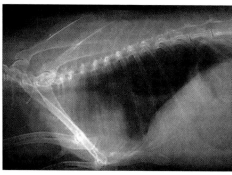 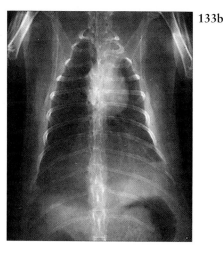

133 The lateral (133a) and a ventrodorsal (133b) views of the thorax of an adult rabbit.
a Review the anatomy of the thoracic viscera of the rabbit.
b What are the abnormalities seen in these radiographs?

134 Three-year-old male rat siblings are housed in a large tank. Two of the rats are in good body condition, but the third is obese. The obese rat has alopecia of the left forelimb.
a What are your differential diagnoses for the alopecia?
b What investigations would help to make a diagnosis in this case?
c How would you manage the condition?

115

132

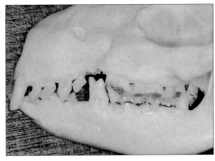

132 a $\begin{array}{cccc} I\underline{3} & C\underline{1} & P\underline{4} & M\underline{3} \\ 3 & 1 & 4 & 3 \end{array}$

b Treat fractured teeth in the same manner as for carnivores. Smooth and polish minor surface damage, but if there is pulp exposure extract the affected teeth. Endodontic treatment is theoretically possible but impractical in such small teeth.

c Reduce the fractures and check occlusion. If there are sufficient teeth on either side of the fracture line to provide anchorage, form intraoral acrylic splints *in situ* in the mouth. As the acrylic is setting, check that normal occlusion is maintained. Splints can be carefully broken away from the teeth once the jaw has healed. As an alternative approach, use enamel bonded composite restorative material between intact rostral teeth to fix the mandible and maxilla together until the fractures have healed (**132**). Fix the rostral mandible in a slightly open position while still maintaining occlusion. The natural gap between the maxillary first incisors, plus the slightly open jaw position, provide space for the tongue to function, reducing the need for hand feeding. Carefully remove the composite material and polish the teeth once the fractures have healed. Release the animal back into its natural habitat if the jaw functions adequately and all other injuries have healed.

133 a The thymus is proximal and ventral to the heart extending to the thoracic inlet. The esophagus, aorta, pulmonary vessels and lymph nodes lie in the mediastinal space dorsal to the heart. The heart is located between the 4th and 6th pair of ribs and lies to the left of midline. The lungs are divided into cranial, middle and caudal lobes. The left cranial lobe is much smaller than the right cranial lobe. The right caudal lobe is subdivided into lateral and medial lobes.
b The radiographs show an interstitial pattern. Hyperinflated lungs are seen on the lateral view but not on the ventrodorsal view. The radiopaque area noted on the ventrodorsal view is the diaphragm which is overlapped by a hyperinflated lung. The pathological diagnosis was emphysema with chronic interstitial fibrosis. The condition was caused by a previous episode of aspiration pneumonia.

134 a Dermatophytosis, ectoparasites, self-trauma and barbering by cagemates.
b The history rules out dermatophytosis and ectoparasites as causes of the alopecia, because only one of the cagemates is affected. Examine hairs plucked from the affected area mounted in mineral oil or place hairs in dermatophyte test media to rule out dermatophytosis. Hairs with broken ends are indicative of trauma. It is important to carefully observe the animals in their own habitat over a period of time. In this case, the two healthy rats were restraining their brother and chewing the fur from his leg.
c Management of behavioral problems such as this are difficult. The only course of action is to remove one of the rats. However, the social dynamics here are complex and the results of separation are unpredictable. In this case, the solution revealed itself. One of the fitter rats died suddenly of respiratory disease some weeks later. After this event, the obese rat lost weight and the hair on its leg grew back.

135 A litter of six Virginia opossums is hand-raised on canine milk substitute. Since weaning three months ago, they have been offered grapes, banana, apple and canned dog food in separate containers. This animal (135) is not walking correctly and is depressed and weak. On physical examination, the long bones are malformed and soft.
a What is your diagnosis?
b How did this condition develop?
c How would you treat it?

136 The owner of a three-year-old rabbit notices tiny white objects moving about in the pet's fur. The rabbit has no other clinical signs. A parasite was found in coat brushings (136).
a What is this parasite?
b Where is the parasite usually found?
c What pathological significance may be attached to finding these parasites in coat brushings?
d How would you control the parasites?

137 A ferret is diagnosed with canine distemper virus. Two other ferrets in the house were exposed to this ferret but are not showing signs of illness.
a How would you treat the disease and what is the prognosis?
b What would you tell the owner about the other ferrets?
c How would you prevent the disease?

135 a Metabolic bone disease due to insufficient calcium in the diet. Confirm the diagnosis with radiographs to reveal demineralized and malformed bones. Opossums naturally move slowly and are not climbers or jumpers, therefore pathologic fractures are uncommon. Collect blood from the lateral tail or cephalic vein to obtain a serum calcium concentration. Normal serum calcium in the opossum is 2.4–2.8 mmol/l (9.6–11.2 mg/dl).

b Opossums often prefer eating fruits and fulfill their caloric requirements with these foods rather than eating the more nutritionally balanced dog food. Fruits have a poor calcium to phosphorus ratio. Others in the cage may be subclinically affected and should be examined thoroughly.

c By correcting the diet. An appropriate diet for the Virginia opossum consists of a non-oily canned adult-formula dog food supplemented with a variety of vegetables, whole mice and hard-boiled eggs. Administer both parenteral and oral calcium supplementation and limit exercise. Place a lateral splint on fractured long bones. Severely affected animals respond poorly to treatment and may not be releasable due to permanent bone deformities.

136 a The mite *Listrophorus gibbus*.

b This is a pelage or fur mite, and it lives within the coat.

c *L. gibbus* is rarely associated with clinical disease, although some instances of apparent hypersensitivity to the mites have been recorded.

d As pelage mites these parasites are unresponsive to ivermectin therapy. Control infestations using topical preparations containing selenium sulfide, lime sulfur or pyrethrins. Clean and treat the environment and treat all the rabbits that are in contact with this patient. Exercise great caution when using insecticides in rabbits. Use powders as opposed to dips unless more aggressive treatment is needed. Avoid the use of insecticides in anorectic, debilitated or obese animals with potential hepatic disease. In the absence of clinical disease it is doubtful whether therapy is justified. However, many owners are uncomfortable with the idea of leaving the mites on their pets once the condition has been brought to their attention.

137 a There is no treatment for canine distemper virus infection. Aim treatment at the secondary bacterial infections and give supportive care. The prognosis for this disease is extremely grave. It is usually fatal.

b The other ferrets, if they were exposed to this ferret, are at risk for developing canine distemper virus. If they have been properly vaccinated against canine distemper virus within the last year, it is highly unlikely they will become infected. Viral transmission is most commonly accomplished by aerosol exposure. Direct contact with conjunctival and nasal exudates, urine, feces and the skin also cause infection. Fomites are implicated in transmission. If the other ferrets are infected, disease will be apparent 7–14 days after exposure.

c By vaccination against canine distemper virus. Vaccinate ferrets with a vaccine approved for ferrets. To impart long-term protection, use a modified live virus. The vaccine should be derived from non-canine cell lines. Otherwise there exists the possibility that the ferret will contract canine distemper virus from a canine cell line vaccine. Vaccinate ferrets every two to three weeks starting at age six weeks for a total of three vaccinations until they are 16 weeks of age. Also, if the ferret has no vaccination history, it is necessary to give it two boosters two to three weeks apart the first time even if it is over 16 weeks of age. Give boosters annually.

138a

138 Two squirrels were hand-raised as part of a rehabilitation program. After a week in an outdoor dirt-floored pen, they exhibit ataxia, tremors and torticollis. A year ago, the pen had previously housed a young raccoon. A recent soil sample from the pen is examined by sedimentation and reveals larvated parasite eggs measuring $63–70 \times 53–58$ µm (138a).
a What is the most likely cause for the squirrel's disease?
b How would you confirm your diagnosis?
c Can this disease be controlled?

139 This clinically depressed young adult raccoon is exhibiting ocular and nasal discharge and has moist rales (139). This wild animal is very easy to handle. It was found in a suburban backyard in an area of the US where there is currently a raccoon rabies epizootic. The captor notices a puncture wound on her hand; however, she does not think that she was bitten by this raccoon.
a What is your diagnosis?
b What course of action should you follow in this case?

139

138b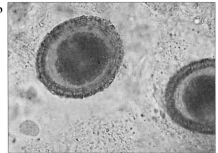

138 a Cerebral larva migrans caused by *Baylisascaris procyonis* larvae is the most likely diagnosis. The eggs are viable in the environment for years and were probably deposited by the raccoon that was in the pen previously.
b The diagnosis is confirmed at necropsy. Examine the thorax and abdomen for larval granulomas. Examine brain sections for larvae and characteristic malacic inflammatory lesions. Place a portion of the brain in two layers of cheesecloth and suspend in a conical-shaped tube filled with warm saline for approximately 12–24 hours. Examine the sediment microscopically for *Baylisascaris* larvae.
c Control of this disease is difficult and it includes environmental decontamination. The soil remains contaminated with viable eggs for years. Eggs can be transported to the surface during gardening or by earthworms and insects. Bleach has little to no effect on the eggs' viability. Treat small areas of contamination on resistant surfaces with a 1:1 mixture of xylene:ethanol. Autoclave or flame metal cages and small items. Flame large areas of soil or concrete with a portable flame gun. Turn, rake and flame the soil several times. Treat raccoons with either pyrantel pamoate (10 mg/kg PO), fenbendazole (50 mg/kg PO q 24 hours for 3 days), or mebendazole (25–40 mg/kg PO q 24 hours for 3–5 days). *B. procyonis* eggs (**138b**) resemble those of *Toxocara* spp. After treatment, burn feces with expelled worms. Treat raccoons weekly or biweekly for four weeks. Because of the zoonotic potential of *B. procyonis*, advise clients to discourage the presence of raccoons into their yards and homes.

139 a The most likely diagnosis is canine distemper. This is a common disease of wild raccoons in the USA. Diarrhea, neurologic signs and hyperkeratosis of the pads can be present in advanced cases of canine distemper in raccoons. To confirm the diagnosis, obtain conjunctival swabs for IFA testing or cytology to identify inclusion bodies. Although less likely, rabies is also considered in this case. The neurological signs of rabies and canine distemper in raccoons are indistinguishable and include paresis, paralysis, self-mutilation, head tilt, facial twitches, circling and lack of fear of humans. Raccoons may be asymptomatic in the early stages of a rabies infection.
b The raccoon came from a rabies endemic area and the captor was possibly bitten, therefore the safest course of action is euthanasia with submission of the head for rabies examination.

140 'Rats explode in laboratory'. This headline was seen recently in a British newspaper regarding an incident involving anesthetized rats.
a What chemical may have caused the explosion?
b At what point in the proceedings might the explosion have occurred?
c What steps would you take to prevent the recurrence of an incident such as this?

141 A two-year-old intact female chinchilla has areas of partial alopecia on the lateral aspects of the limbs of four weeks duration. She is caged alone and her owner has just started full-time employment.
a What are your differential diagnoses for this case?
b How would you confirm the diagnosis?
c What are the predisposing factors to self-trauma in the chinchilla?

142

142 Incisor malocclusion is a common problem in pet rabbits (**142**). Overgrowth can be managed by trimming, but this procedure usually needs to be repeated at intervals throughout the patient's life.
a What is an alternative method of managing non-functional incisor teeth?
b What investigations would be most useful in assessing the prognosis for this procedure?
c What dietary changes would you recommend for rabbits that have no incisor teeth?

140 a The explosion was caused by the use of anesthetic ether. Take great care when using anesthetic ether to avoid generating sparks or excessive heat. This includes the use of electrosurgery, laser surgery or sparks from any electrical equipment.
b In this case, the rats were killed using ether and immediately placed in a freezer. This resulted in a build-up of ether fumes. When the freezer motor switched on, the spark was sufficient to cause an explosion.
c Use a ventilated, spark-proofed fume cupboard prior to placing ether-killed animals in the freezer. Although ether has been traditionally utilized in research facilities and by some practitioners to reduce costs, the risk of explosion now severely limits its use. In addition, ether causes an unpleasant irritation of the mucous membranes.

Other causes of explosions include dropped oxygen cylinders, valves of oxygen cylinders contaminated with grease resulting in spontaneous ignition of the high-pressure oil/oxygen aerosol and the use of electrosurgery on tissue that has been prepared with alcohol.

141 a Dermatophytosis, fur slip and self-barbering.
b Microscopically examine epilated or plucked hairs from the affected area. The distal tips of the hairs will be broken and damaged in cases of fur chewing. With fur slip, the hairs appear normal because almost complete shedding occurs. Plucked hair from dermatophytosis cases may be frayed or damaged. Mount the hairs either in mineral oil or KOH to assess for fungal hyphae or spores.
c The predisposing factors include stress, chronic disease, malnutrition and the absence of dust baths. Ectoparasite infestation is uncommon in chinchillas. Examples of stress include gestation, cagemate death or disease, overcrowding or moving into a new cage or location. There is also a theory that in some undiagnosed cases there may be an inherited behavioral trait or endocrine disease. This chinchilla was barbering due to the change in the household schedule caused by the owner going back to work.

142 a Extraction.
b Use serum biochemistries and hematology to detect concurrent systemic disease. Thoroughly examine and assess the condition of the cheek teeth. Clinical evaluation with palpation and visual examination, even under anesthesia, will only reveal about 50% of the dental or oral lesions present. Radiography is an essential part of the workup in these cases because it provides information about the tooth roots. Early signs of cheek tooth overgrowth or root extension may be reversible, but if gross changes are seen on radiographs, the prognosis is poor. Check for incisor tooth root deformity. Any enlargement of angulation towards the root apex will make extraction difficult.
c The dietary requirements following incisor extraction are basically the same as for all rabbits, which includes good quality hay and fresh herbage. Cut bulky items, such as carrot or celery, into thin strips so they can be picked up by the prehensile lips and fit into the rabbit's mouth. The tongue is then used to move the material to the cheek teeth for grinding. Most rabbits can graze normally and can manage long hay strands easily. Advise owners to brush the coat frequently to remove dead hair because the loss of incisors prevents the rabbit from performing this grooming procedure.

143 With regard to the rabbit in question **142**:
a How would you extract incisor teeth in the rabbit?
b What is the most common complication during incisor extraction and how would you manage it?

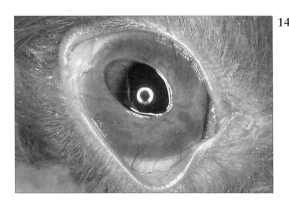

144

144 This photograph (**144**) shows the right eye of a four-month-old female rabbit which developed this progressive growth over the cornea in the last two weeks. The rabbit appears otherwise healthy and the left eye is normal.
a What is your diagnosis?
b What is the cause of this condition?
c How would you treat it?

143 a Examine the pre-operative lateral radiograph to assess the length and curvature of the tooth roots. Anesthetize the patient and trim the main incisors so that 5–10 mm of the crowns are visible. Treat any cheek tooth irregularities and oral mucosal lesions at this time. Extract the peg teeth first because they are often fractured during extraction of the main maxillary incisors. Clean the gingiva with an antiseptic solution. Work a fine carnivore-type dental elevator around each peg tooth to sever carefully its periodontal ligaments. Avoid leverage on the tooth crowns because this can lead to root fracture. After the teeth are loosened, lift them out of their sockets using fingers or a small pair of forceps. The marked curvature of the main incisors makes elevation of these teeth more difficult. Use a number 11 scalpel blade to separate the gingival attachments and the medial and lateral periodontal attachments. Insert a fine dental elevator medial to each of the teeth and apply slight lateral pressure for a minute or two. Repeat this several times until the tooth is loosened. Grip the crown of the tooth with fingers or forceps and rotate the tooth out of its socket along the line of eruption. The tooth sockets usually fill with blood which clots rapidly. If hemorrhage persists or if the gingiva was torn, close the wounds with a fine absorbable suture material. Culture the tooth socket and administer antibiotics in cases where tooth root infection is suspected. Use analgesia postoperatively. Provide food and water as soon as the patient is ambulatory. Do not use soft food because it adheres to the wound area.
b For the novice, root fracture is quite common, particularly affecting the peg teeth. Forewarn the owner of this possibility. The best way to manage these cases is to extract the remaining incisors leaving the fractured root in place. Within six to eight weeks the tooth has usually regrown and is then extracted at a second procedure. If the tooth does not regrow, take radiographs to check for impaction. If this has occured, perform an osteotomy to expose and remove the root remnant.

144 a Corneal membranous occlusion, or centripetal conjunctival migration, or pseudopterygium. The disorder is usually progressive, may be unilateral or bilateral and, in extreme cases, results in blindness due to occlusion of the pupil.
b Unknown, but it may represent a congenital symblepharon; however, it is also known to occur in adult rabbits. The architecture of the membrane may be quite complex and is usually lined with conjunctival epithelium on both surfaces. Frequently, the membrane adheres only to the limbus and not the cornea. Rarely, loose to moderately firm, focal adhesions to the peripheral cornea occur. No evidence of inflammation is present in this eye and microbiologic cultures are negative for *Mycoplasma*, *Chlamydia*, and *Pasteurella* spp. The Schirmer tear test is also normal at 8 mm/minute (the normal mean + SD Schirmer tear test for the rabbit is 5 + 3 mm/min) suggesting a tear film disorder is not involved. Deeper intraocular structures are normal.
c Excise the membrane. Some rabbits require a superficial keratectomy if there are focal adhesions to the cornea. Preserve a small, perilimbal ring of conjunctival epithelium as this is where the progenitor cells of the cornea are located. Recurrence following excision is common. To prevent regrowth, anchor the cut edge of the bulbar conjunctiva to the episclera at least 2 mm away from the limbus with a few, fine absorbable sutures. Administration of topical corticosteroids such as a 0.1% dexamethasone and polymyxin B/neomycin combination three to four times daily for 10 days postoperatively may reduce the possibility of membrane regrowth. Use steroids in rabbits with great caution. Use a broad-spectrum topical ophthalmic antibiotic three to four times daily until the corneal defect has healed.

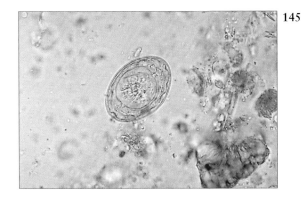

145 These eggs (30–50 µm) are found on a fecal floatation from an anorectic, depressed hamster (**145**). Internal hooks and filamentous processes are evident.
a Identify the eggs.
b Is this organism pathogenic for hamsters?
c How was the hamster infected?
d Is this organism zoonotic?
e How would you treat and control this parasitic infection?

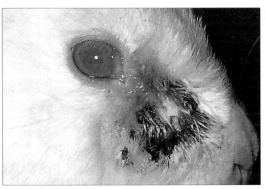

146 A pet rabbit has a unilateral facial dermatitis (**146**). The lesion has become progressively larger over the past two weeks. The rabbit appears otherwise normal and does not appear to be in any discomfort.
a What is the etiology of the lesion on this rabbit's cheek?
b How would you manage this condition?
c What is the procedure for cannulating the lacrimal duct in the rabbit?

145 a Eggs of the tapeworm *Hymenolepis nana*.
b In large numbers, this organism causes intestinal occlusion, impaction and death in hamsters.
c This parasite has both a direct and an indirect life cycle. The adult tapeworms live in the small intestine. In the direct life cycle, gravid proglottids detach from the body and disintegrate in the intestine releasing eggs which are immediately infective when passed in the feces. Hamsters ingest these eggs which hatch in the small intestine and the onchosperes (embryos) penetrate into the mucosa and lie in the lymph channels of the villi. They develop into metacestodes (cysticercoids) which emerge from the villi in five or six days, attach to the intestine and mature into adult tapeworms. In the indirect life cycle, the eggs are ingested by insects (grain beetles, fleas, cockroaches) and develop into the next larval stage (cysticercoid/metacestode). The hamster is infected by ingesting these insects. Finally, autoinfection is also possible. Some of the eggs that are released when the proglottids disintegrate in the intestine may hatch in the hamster and the onchosperes penetrate the intestinal villi. Cysticercoids emerge in five or six days and develop into adult tapeworms without ever leaving the hamster.
d *H. nana* infects rodents and humans. Humans are infected by ingesting the eggs or the insect intermediate hosts. Autoinfection also occurs. Infection in humans is inapparent to mild, but if massive numbers of worms are present, damage to the intestinal mucosa produces enteritis. Only a moderate number of parasites are needed to produce diarrhea, anorexia and abdominal pain in children.
e Praziquantel (5 mg/kg SC or 10 mg/kg PO q 14 days for two treatments) is the drug of choice as it is active against both the larval and adult stages. Incinerate bedding and feces from the hamster to prevent the spread of the eggs. Clean and sterilize or replace the hamster cage. Break the life cycle by sanitation and insect control. Rodent control is essential since wild rodents may also be a source of contamination.

146 a Tear overflow or epiphora, caused by an obstructed nasolacrimal duct. The nasolacrimal duct of the rabbit is unusual in that it has two sharp bends along its course. A build-up of purulent material at these duct deviations leads to obstruction. In addition, these ducts pass close to the root apices of the premolars and incisors and may become obstructed if there is tooth root disease.
b The best chance for a complete cure lies in aggressive therapy early in the course of the disease. Perform skull radiographs to evaluate for dental disease. Clear the duct by cannulation and irrigation with a sterile saline solution. Collect a few drops of the flushed secretions exiting the nares by free catch on to a culturette for bacterial culture and sensitivity. Use appropriate antibiotics both topically and systemically for several weeks. Recheck the rabbit and flush the tear ducts with an antiseptic or antibiotic solution at seven-to-ten day intervals until the condition resolves. Give a guarded prognosis for a complete cure because in some rabbits the condition will recur despite aggressive therapy.
c There is only one nasolacrimal cannaliculus and punctum in the rabbit located in the lower lid. Achieve lacrimal duct cannulation without general anesthesia. Use a local ophthalmic anesthetic to desensitize the eye. To visualize the punctum, either pull the lower lid down and roll it outward or grasp the lid edge and gently pull it laterally. Cannulation can be challenging in the presence of a persistent discharge and severe inflammation of the peripunctal area. Apply pressure below the punctum to cause the lips of the punctum to 'pout' allowing better access to the duct.

147 An owner requests a stool culture on his clinically normal African hedgehog (147). An infant in the household recently developed an acute illness with fever accompanied by diarrhea. The infant was breast-fed and had no direct contact with the hedgehog.
a What organism would you suspect and what tests would you carry out?
b What protocols minimize the zoonotic risk of owning an African hedgehog?

148 A three-month-old hamster is presented for a routine health examination. The food dish contains a sunflower-based seed diet.
a What deficiencies and excesses exist with an all-seed diet?
b What recommendations would you make to correct the diet?

149 A six-month-old male guinea pig has alopecia accompanied by crusting and flaking skin (149). The patient appears to be intensely pruritic and uncomfortable.
a What is your diagnosis?
b How would you confirm the diagnosis?
c Is this condition contagious to other pets?
d How would you treat this condition?

147 a The illness of the child is consistent with salmonella infection. *Salmonella* serotype *Tilene* has been documented by the National Center for Infectious Diseases, CDC, USA, in two human cases in which the African hedgehog has been implicated as the source of the infection. The incidence of *Salmonella* spp. in domestically raised African hedgehogs is not known. It is important to specifically request both a *Salmonella* culture and *Salmonella* serotyping. Most laboratories will not culture for *Salmonella* unless specifically asked to do so.
b Reduce the risk of transmission of *Salmonella* spp. from an infected pet to household members by washing hands thoroughly after handling the pet. This is particularly important before eating or handling food. Avoid contact with the pet's feces. Perform microbiologic cultures on pet hedgehogs for *Salmonella* spp. at the time of their annual examination or postpurchase examination if there is concern for zoonotic disease; however, some species of *Salmonella* are non-pathogenic and a negative culture does not necessarily mean that the pet is free of *Salmonella*. The effectiveness of antibiotic treatment for the clearance of *Salmonella* has yet to be evaluated in hedgehogs.

148 a A sunflower seed-based diet is deficient in calcium, protein, vitamin A, vitamin D, riboflavin and vitamin B$_{12}$. A lack of sufficient protein can result in alopecia. This type of diet also contains excessive amounts of fat (30% or higher) which can lead to a functional vitamin E deficiency as well as obesity.
b Hamsters are granivorous by nature. In the wild, they eat a variety of plants, seeds, fruit and some insects. In captivity, the hamster should be fed a pelleted rodent diet that contains a minimum of 16% protein and a maximum of 5% fat. Avoid seed diets or high-fat treat foods. For variety, up to 10% of the diet can consist of whole grains, bread, fresh fruit and green leafy vegetables. Hamsters will take the food and store it in various places in the cage, therefore feed perishable foods only in amounts that will be consumed in one day to avoid spoilage. Hamsters can obtain water from sipper bottles or shallow bowls that are secured to the cage.

149 a The likely diagnosis is infestation of the sarcoptiform mite, *Trixacarus caviae*. The sarcoptic mite of dogs, *Sarcoptes scabiei*, can also affect the guinea pig in a similar manner.
b Skin scrapings of affected areas will reveal the mite in its various life stages.
c The *T. caviae* mite is usually only contagious to other guinea pigs, but there have been reports of it causing urticaria in humans. *S. scabiei* may affect a variety of species, including humans.
d Administer ivermectin (0.2–0.4 mg/kg SC or PO). Repeat this dose two more times at 7–10 day intervals. This treatment is effective for the eradication of either *T. caviae* or *S. scabiei*. Treat cagemates and have the owner thoroughly clean the cage and cage furniture after each treatment. Advise the owner of the potential for zoonotic disease in the case of *S. scabiei*.

Although it is usually unnecessary to provide topical therapy for these cases because ivermectin is very effective, some patients are in extreme discomfort from the intense pruritus or have a secondary bacterial dermatitis. In these cases, gently soften and remove the majority of skin crusts with a warm bath. Sparingly apply a topical antibiotic preparation appropriate for guinea pigs.

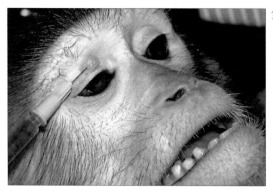

150 A two-year-old male long-tailed macaque requires a physical examination, vaccinations and evaluation of its herpes B virus status. The macaque is immobilized with ketamine. The physical examination is normal.
a What is the purpose of the injection shown (**150**) and what substance is being injected?
b Which vaccinations would you recommend?
c What would you tell the caretaker regarding these vaccines?
d What test(s) should determine the herpes B virus status of the monkey?
e What does a positive test mean?
f Is herpes B virus zoonotic?

151 A group of hand-raised, recently weaned grey squirrels live together in a large cage at a wildlife rehabilitation center in the USA. This squirrel and some of the others in the group have raised, flattened, sparsely haired nodules ranging in size from 0.5–2 cm in diameter on their faces and bodies (**151**).
a What is your diagnosis?
b What can you do for the affected animals?
c What can be done to prevent disease transmission?
d What other species can be affected?

150 a This is the intradermal subpalpebral injection site used for tuberculosis hypersensitivity testing. Old mammalian tuberculin is the recommended antigen.

b Tetanus toxoid and measles virus vaccination.

c There are no vaccines labelled for use in primates, therefore it is an off-label use of a drug. The measles vaccine is only available as a modified live vaccine and the animal actually sheds the vaccine virus following vaccination. It is estimated that the shedding period is 3–10 days post-inoculation. During this time, isolate the macaque from infants under the age of 18 months, elderly people and people who have never been vaccinated or exposed to the disease. Also, measles vaccination is immunosuppressive to primates.

d Serologic testing is available for herpes B virus (at Southwest Foundation for Biomedical Research, PO Box 28147, San Antonio, TX 78254, 210-673-3269).

e Evidence of exposure to the virus but not necessarily infection. A negative result only means there were no circulating antibodies to the virus at the time of testing. Since an animal can be infected with herpes B virus and be seropositive or seronegative at different times, a negative result has no meaning on the herpes B virus status of the animal. Oral pharynx lesions can be cultured for the virus. Positive culture indicates that the macaque is shedding virus from the oral cavity and is infected. Negative culture results are more common and only indicate that no virus was cultured at the time of sampling. Since herpes B virus is a latent virus, an animal is considered infected for life.

f Consider all macaques potential carriers of herpes B virus regardless of test results. Viral transmission from macaque to humans is accomplished via bite wounds, scratches, exposure of human mucous membranes to fluids or secretions from the monkey, and infected syringe needles. Herpes B virus can be a fatal infection to humans. Ideally, do not permit contact between macaques and humans.

151 a Squirrel fibroma caused by a poxvirus. This is the most common cause of nodular skin disease of squirrels in North America. Use cytology or histopathology to differentiate this disease from neoplasia. Cutaneous *Cuterebra* infestation also appears as multiple raised nodules, but the larva is easily visualized through the central opening into its capsule.

b Affected animals can recover without treatment in four to six weeks. Institute supportive care if they become anorectic. Anecdotal reports claim that the administration of antibiotics and vitamin A aid in recovery due to the control of secondary bacterial infections and support of damaged tissues.

c Isolate the affected squirrels. Institute an external parasite control program to prevent disease transmission by biting insects, such as mosquitoes. Some wildlife rehabilitation facilities recommend euthanizing affected animals as a control measure.

d Poxviruses are relatively species specific, but this particular virus will affect groundhogs and other species of squirrel, such as the fox squirrel and red squirrel.

52a

152b

152 A progressive disease process has effected the left eye of a five-year-old female rabbit (152a). The left eye is now blind and lacks a direct pupillary light reflex (152b).
a What abnormalities are evident?
b What diagnostic evaluations would you consider?
c What are your treatment options?

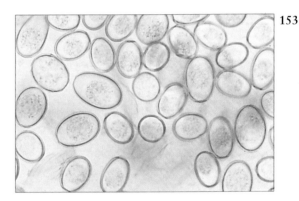
153

153 Several young rabbits in a rabbitry developed anorexia and diarrhea. A fecal flotation is performed and these organisms were recovered (153). Some of the ellipsoidal organisms measured 28–40 × 16–25 μm and had a micropyle (cap/operculum). Many other organisms are seen on the slide including ones without micropyles. What are the organisms and could they be related to the anorexia and diarrhea?

131

152 a The left globe is exophthalmic and deviates ventrally. The third eyelid protrudes and a whitish plaque covers the corneal surface. Purulent material emanates from a tract superior to the globe. These findings indicate a space occupying orbital lesion, such as a retrobulbar abscess, with a septic component. The exophthalmos causes an exposure keratopathy resulting in a whitish corneal plaque. The orbital disease also affects the optic nerve resulting in blindness and the presence of the afferent pupillary deficit.

b Perform a complete physical examination. Obtain a CBC, serum biochemistries and skull radiographs. Perform a cytologic evaluation of the discharge. Obtain a microbiologic culture of the tract. Use ocular and orbital ultrasonic examinations and, if available, it is additionally helpful to use an advanced imaging method, such as a CT scan or MRI, to determine the extent of the pathologic involvement. Do a thorough oral examination to rule out dental disease as a cause of a retrobulbar abscess. Radiograph the dental arcades to identify molar malocclusion or associated dental abscess.

c The prognosis for this condition is guarded, particularly if *Pasteurella* sp. is confirmed as the causative agent. Perform an exenteration (removal of the globe and orbital contents) to remove as much diseased tissue as possible. Treat the patient while it is in the hospital with appropriate systemic antibiotics based on microbiologic culture and sensitivity results and continue antibiotic therapy at home for several weeks. Give a guarded prognosis even with aggressive therapy.

153 The organisms seen are coccidial oocysts representing several species of *Eimeria*. Those measuring 28–40 × 16–25 µm are probably *E. stiedae*. It is an extremely pathogenic protozoan parasite that lives in the bile duct epithelium of the rabbit. In severe infections this parasite produces hepatic dysfunction and may cause death. Rabbits are infected by ingesting infective oocysts from the environment found in the food, water or bedding. Oocysts hatch and release sporozoites that penetrate the intestinal mucosa and travel via the mesenteric lymph nodes and hepatic portal system to the liver. Here they enter the epithelial cells of the bile ducts and liver parenchymal cells. They multiply asexually (schizogony), rupture the cells and each newly released organism invades a new cell and multiplies asexually again. Eventually, asexual multiplication ceases and they undergo sexual reproduction in the cells and produce oocysts that are passed unsporulated through the bile ducts into the intestine. The oocysts pass out with the feces and must sporulate to the infective stage in the environment. This takes at least two days, therefore reinfection cannot take place through the eating of cecotropes. Disease and illness occur during the asexual multiplication stages and the host may die before oocysts are shed in the feces. The prepatent period is 18 days. Unfortunately, it is difficult to distinguish *E. stiedae* from other *Eimeria* spp. that infect the intestinal tract of rabbits. These intestinal coccidia may produce diarrhea, particularly in the young because they multiply in the intestinal epithelium before producing oocysts which pass in the feces.

154 To improve rabbit anesthesia, the use of premedications and injectable anesthetics is recommended.
a What anesthetic premedications are useful in the rabbit and why?
b Ketamine, ketamine/diazepam and ketamine/xylazine are all used for chemical restraint and light anesthesia in the rabbit. What are some limitations of each in the rabbit?
c What problems are associated with the use of barbiturates or tiletamine/zolazepam for anesthesia in the rabbit?

155 During a routine physical examination on a four-year-old male castrate ferret, a large spleen is palpated (**155**). The ferret appears to be otherwise normal.
a What are your differential diagnoses?
b What is the most common cause of a large spleen in a ferret?

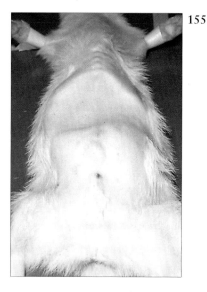
155

154 a Injectable anesthetic agents are useful to sedate rabbits for examination, intravenous catheter placement or before induction with inhalant anesthesia. Handle and restrain a sedated rabbit carefully to prevent injury to its delicate spine and legs. Use tranquilizers such as acepromazine (0.1–1.0 mg/kg IM) in combination with ketamine (20–30 mg/kg IM) to improve the quality of anesthesia produced. Administer diazepam (2–10 mg/kg IM) to provide sedation and muscle relaxation. When diazepam (0.5–1.0 mg/kg IM, IV) is used with ketamine (10–20 mg/kg IM, IV) to improve the quality of anesthesia, enhance the effect by giving diazepam approximately 30 minutes before ketamine. Alternatively, use the sedative analgesic, xylazine (2–5 mg/kg IM, SC) with ketamine (35–50 mg/kg IM). Use xylazine only in healthy animals. Its use is associated with muscle relaxation, peripheral vasoconstriction and decreases in the heart rate, arterial blood pressure and respiratory rate.

Atropine is not routinely used as a premedication in rabbits. Some rabbits have atropinesterase and, therefore, metabolize atropine rapidly. Glycopyrolate may prove to be a more useful anticholinergic agent in rabbits.

b Ketamine (35–50 mg/kg IM) produces sedation and chemical restraint but does not provide enough muscle relaxation and analgesia for most surgical procedures. Ketamine (35–50 mg/kg IM) in combination with xylazine (3–5 mg/kg) provides a variable plane of anesthesia and may require supplemental inhalant anesthesia for surgery. Decreases in blood pressure, respiratory rate, heart rate and body temperature are associated with ketamine and xylazine combinations. Local irritation is reported following intramuscular administration of ketamine/xylazine. To avoid this, divide the total dose and administer at several sites. A combination of ketamine (20–40 mg/kg IM) and diazepam (5–10 mg/kg IM) provides a variable plane of anesthesia. Supplement as needed with an inhalant anesthetic for endotracheal intubation and surgery.

c Barbiturates have potent respiratory depressant effects in rabbits. Severe laryngospasm can also be a problem at light levels of barbiturate anesthesia. Barbiturates have a very narrow margin of safety. Tiletamine/zolazepam can cause renal toxicity in rabbits, with the tiletamine implicated as the causative factor in a dose-dependent manner.

155 a Extramedullary hematopoiesis, neoplasia, infection, hyperplasia, congestion due to cardiac disease and portal hypertension.

b Extramedullary hematopoiesis. The cause of extramedullary hematopoiesis is not known. It is a common, but benign, problem in many ferrets over three years of age. Neoplastic conditions include lymphosarcoma, hemangioma, hemangiosarcoma and metastasis from distant sites. Of these, lymphosarcoma is most common. Infection is a rare cause of splenomegaly although it is commonly seen with Aleutian disease. Splenomegaly due to hyperplasia or portal hypertension has not been reported in the ferret.

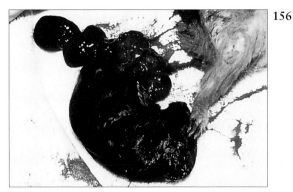

156 A 10-month-old female guinea pig has a fetus protruding from her vagina (**156**). The fetus is dead and the sow is weak and in shock.
a How would you treat the guinea pig?
b What is the cause of the condition?

157 A ferret had been diagnosed with pancreatic beta cell tumors. The owner is trying to decide whether to treat the ferret with medicine or surgery. What do you advise?

158 During the routine annual physical examination of a chinchilla, a substantial amount of fur easily epilates (**158**).
a What is the most likely cause of this hair loss?
b How would you prevent it from happening?

156 a Initially, treat for shock and administer appropriate supportive care. Give an intravenous or intraosseous balanced electrolyte solution with 5% dextrose immediately. Follow this with a rapidly acting corticosteroid, such as methylprednisolone acetate, and antibiotics, intravenously. Correct hypothermia. Delay blood collection until the sow is stable. Gently attempt to manually extract the fetus. Perform an emergency Cesarean section if the fetus is wedged in the birth canal. An assistant retropulses the fetus carefully through the birth canal while the surgeon gently applies traction from within an opened uterus. Perform an ovariohysterectomy at this point if the sow is stable.
b The most common cause of dystocia in sows is the inability of the pubic symphysis to expand under the influence of the hormone relaxin. It then does not separate as needed just prior to parturition. This condition occurs in sows bred for the first time over the age of eight months. In addition, older, obese sows have larger fat pads in their pelvic canals that can impede parturition.

157 Treatment for pancreatic beta cell tumors in ferrets consists of medication, surgery or both. Surgery is primarily a debulking procedure as it rarely removes all of the tumor. Medical treatment is aimed at controlling the signs but does not halt the progression of this disease. Advise the owner that when insulinoma is diagnosed, the patient will require management of this disease for the rest of its life. A recent study revealed that ferrets receiving medical treatment for insulinomas lived an average of six months postdiagnosis. However, some ferrets can live as long as two years on medical therapy alone. Medical treatment most often consists of prednisone and/or diazoxide. The owner should discontinue any sugary treats in the diet. The diet should consist of frequent feedings of a high-quality cat or ferret food. Begin medical treatment when the blood glucose concentration is below 3.66 mmol/l (65 mg/dl). All blood glucose concentrations should be based on a fasting sample. Fast the ferret from four to no more than six hours. If prednisone is given first, administer at 0.10–0.50 mg/kg PO q 12 hours. Begin treatment with a low dose gradually increasing the amount given as dictated by the ferret's signs. Obtain fasting blood glucose samples every one to three months. As the tumor spreads, it becomes increasingly resistant to the effects of the prednisone. The second commonly used medication is diazoxide (5–30 mg/kg PO q 12 hours) which prevents the release of insulin. Administer both medications orally. Start treatment with one medication and as resistance increases, add the other medication to the treatment regimen. There are no contraindications in giving the medications concurrently. Since prednisone increases the workload on the heart, give cautiously in ferrets with heart disease. It is useful to screen ferrets for insulinoma by obtaining routine fasting blood glucose samples every six months once the ferret reaches three years of age. Early detection and consequent treatment improves the long-term prognosis.

158 a Chinchillas are generally easy to handle but can be nervous. Since they rarely bite, head restraint is not necessary. However, rough handling, scruffing or quick grasps at a fleeing animal may result in a condition called 'fur slip'.
b Wear light cotton or clean latex gloves when handling chinchillas to prevent hand oils from damaging the delicate fur. Handle chinchillas by gently cupping the entire animal in both hands, always supporting the rear quarters. If necessary, grasp the base of the tail to stabilize the pet before cupping the free hand around the body. Although 'fur slip' is a cosmetic problem, it can damage a pelt and may give the owner a poor impression of the veterinarian.

159 A seven-year-old intact female rabbit has a bilateral rear limb paresis and inappetence progressing over a two-week period. The rabbit maintains a splay-legged position, is weak on all four limbs and is reluctant to stand. Spinal reflexes are normal in all four limbs. The ventrum and perineum are superficially ulcerated by constant contact with urine.
a What are your differential diagnoses for generalized weakness in the rabbit?
b What are your differential diagnoses for chronic staining of the perineum?
c What diagnostic tests would you carry out?

160 A two-year-old female rabbit is anorectic for three days. The rabbit is fed a diet consisting primarily of pellets with hay and grain-based treats given on occasion. The owner reports that the rabbit refused to eat pellets for two days, but continued to eat hay, treats and the newspaper bedding in the cage before becoming anorectic. The rabbit has produced scant, dry feces since the onset of anorexia. It has a small amount of palpable intestinal gas, but is otherwise normal in appearance and attitude. An abdominal radiograph is taken (160). What is your diagnosis?

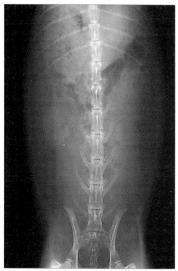
160

159 a Metabolic diseases (i.e. renal failure and hepatic), neoplasia, ingested toxins, mucoid enteropathy, severe infections, cerebrovascular accident, trauma, encephalitis damage due to *Encephalitozoon* (*Nosema*) *cuniculi* or *Baylisascaris* spp., vertebral spondylosis, spinal trauma or neoplasia. In this case, the apparent rear limb paresis is due to generalized weakness caused by metabolic disease.

b Cystitis, cystic calculi, reproductive disease (i.e. pyometra, uterine neoplasia), renal disease, rabbit syphilis, obesity (perineal skin folds around the perineum trap urine), urinary incontinence, spinal disease, CNS disease, systemic disease resulting in generalized weakness and disease causing painful movement (i.e. arthritis, podo-dermatitis).

c Hematology, serum biochemistries, urinalysis and abdominal radiographs. In this case, CBC abnormalities included a non-regenerative anemia and hypoproteinemia. The serum chemistry profile demonstrated an increased BUN and creatinine. A low specific gravity was the only abnormality on urinalysis, indicating a diagnosis of renal failure. It is important to perform a complete physical examination on all rabbits presenting with rear limb paresis to differentiate neurologic disease from generalized weakness.

160 The most likely diagnosis is GI stasis. This is often the end result of GI hypomotility caused by a diet low in fiber and/or high in starch. When GI motility is slowed, the stomach and the cecum do not empty properly and the material they contain may become partially dehydrated. When the patient becomes anorectic, complete GI stasis occurs resulting in further dehydration and impaction of the stomach and cecal contents. Often there is a 'halo' of air around the gastric contents seen on radiographs. This is frequently what is occurring when rabbits are diagnosed with a gastric trichobezoar or 'wool block'. The term 'trichobezoar' is misleading because the problem is not exclusively an accumulation of hair. Nearly every rabbit has hair present in the gastric contents due to continual grooming. When the gastric contents become dehydrated the hair, which represents only a portion of the stomach contents, is bound together in a matrix of compacted ingesta. This is different than a true trichobezoar which is a tightly compacted mass composed of nearly 100% hair, as seen in cats or ferrets.

It is typical to observe that rabbits crave high-fiber items, such as hay and paper, when they experience GI motility problems. Feces decrease in size and eventually cease altogether when the rabbit becomes anorectic. Early in the disease, these patients will have little or no abdominal discomfort and will appear bright and alert. Anorexia can eventually lead to hepatic lipidosis which will result in further deterioration of the patient's condition. As the disease progresses large quantities of painful gas build up through the entire GI tract.

GI obstructive disease does occur in rabbits. The most common causes include rubber, plastic, small masses of dried ingesta mixed with hair or carpet fibers and stricture of the GI tract from post-surgical adhesions. These patients are acutely and severely depressed and exhibit abdominal discomfort. Radiography in these cases reveals significant dilation of the stomach.

161 With regard to the rabbit in **160**:
a What are the major contributing factors to this condition?
b How would you treat this rabbit?

162 A four-month-old ferret is found outside. It
is weak, tachypneic and sneezing, and has a co-
pious mucoid nasal discharge, has chin swelling
and crusting and thickened foot pads (**162**).
The ferret is brought into a house where there
are two other ferrets. Despite supportive care,
the ferret develops ataxia and torticollis and is
now comatose.
a What are your differential diagnoses?
b What is your diagnosis?
c What tests would you perform?

162

161 a The single most common contributing factor in GI motility problems in the rabbit is low dietary fiber. The non-digestible fiber portion of the diet is essential to promote normal GI tract motility. In addition, a high starch diet, ingestion of toxins or inappropriate antibiotics can lead to changes in the cecal pH resulting in dysbiosis which in turn can cause cecal impaction and lowered GI motility.

b Direct therapy towards restoring normal GI tract motility and correcting underlying dietary problems. Rehydrate the stomach contents and stimulate GI motility through administration of a high moisture and fiber diet, parenteral fluids and GI motility drugs. Offer grass hay and fresh leafy greens, such as dandelion greens, parsley, romaine lettuce, carrot top or kale. Many rabbits will eat these high fiber foods eagerly avoiding the need for syringe feeding. If the rabbit refuses to eat on its own, administer appropriate assist feedings. Avoid the use of supplements that are high in fat or starch. Avoid the use of psyllium powder or other fiber products that draw water out of the colon. Due to the nature of the physiology of the proximal colon, the use of these products may lead to cecal or colonic impaction. Typically stools may not be produced for up to 72 hours after the institution of assist feeding. After replacing fluid deficits, administer lactated Ringer's solution at a maintenance level of 75–100 mg/kg/day SC, IV or IO. Administer drugs to promote GI tract motility, such as metaclopramide (0.2–1.0 mg/kg SC or PO q 12 hours), or cisapride (0.5–1.0 mg/kg PO q 12–24 hours) until normal stools are produced. Analgesics, such as flunixin meglumine, may be helpful initially to reduce pain from excessive GI gas. Surgery is rarely indicated in these cases. For prevention, recommend a diet of free choice quality grass hay, fresh herbage and limited concentrate foods. Avoid high starch treats.

162 a Canine distemper virus, influenza virus, bacterial pneumonia or mycotic pneumonia.

b Canine distemper virus infection is the most likely diagnosis. Although the other diseases can cause severe respiratory infection, only canine distemper virus causes both respiratory and neurologic disease signs. Canine distemper virus also causes chin swelling and crusting which are cardinal signs of this disease and appear before thickened foot pads are apparent. Swelling and crusting may also be present in the perineal and inguinal area of ferrets infected with canine distemper virus. This ferret was found outside where it likely came in contact with infected animals and the vaccine status of this ferret is unknown. If only neurologic signs were apparent, rabies virus infection and listeriosis should be considered in the differential diagnosis although these are rare diseases in ferrets.

c Perform a fluorescent antibody test for canine distemper virus antigen on conjunctival smears, mucous membrane scrapings or blood smears to diagnose canine distemper virus. This is useful only in the first few days of disease. The modified live viral strains used for vaccination do not interfere with this testing method. To aid in the diagnosis of canine distemper virus infection, perform hematology. As with other viral infections, a leukopenia can be observed. Radiographs characterize the extent of the pneumonia. Perform a lung wash to determine the cause of the pneumonia. A secondary bacterial pneumonia is usually present which can be responsible for the respiratory signs. Perform cytology on a cerebrospinal fluid tap to determine the cause of the neurologic signs. Diagnose this disease definitively at postmortem. Canine distemper virus inclusion bodies are generally found in the epithelial cells of the trachea, urinary bladder, skin, GI tract, lymph nodes, spleen and salivary glands.

163

163 A two-year-old male guinea pig is anorectic and reluctant to walk (**163**). Pain is elicited during joint palpation and it moves with a stiff gait. The incisors appear normal, but the guinea pig has difficulty prehending its food. The skin has petechial hemorrhages present on the right side of the neck and over the tarsal-metatarsal joints.
a What is your diagnosis?
b What aspect of the husbandry is crucial in establishing a diagnosis ?

164 A four-year-old spayed female ferret is found unable to walk. On physical examination, there is no voluntary motor movement in the hindlimbs. All reflexes are assessed as normal and muscle atrophy is not apparent.
a What are your differential diagnoses?
b What tests could you performed to determine the cause of this condition?

163 a Hypovitaminosis C, also known as scurvy. Guinea pigs, like humans, do not produce their own vitamin C and require a dietary source. The minimum daily requirement for vitamin C in the guinea pig is 10 mg/kg/day in the non-breeding animal and 30 mg/kg/day in the pregnant and lactating sow. Signs of vitamin C deficiency are consistent with damage to connective tissue, and include swollen painful joints and costochondral junctions resulting in reluctance to move, crying in pain when touched, poor coat condition, dysphagia and hemorrhage around joints, muscles and skin. The angle of the teeth may change as they become loosened, resulting in malocclused and overgrown molars and the resultant inability to prehend food. Some guinea pigs present with chronic mild upper respiratory signs and increased ocular secretions, while others may present with diarrhea and inappetence.
b The diet is a crucial factor in this case. Owners are either unaware of the vitamin C requirement of guinea pigs, or believe that commercial guinea pig pellets are an adequate diet. Vitamin C is a water soluble vitamin and its potency decreases rapidly in the feed, particularly if exposed to excessive heat or moisture. Most commercial guinea pig pellets lose their vitamin C potency within 90 days after milling, therefore it is important to determine the age of the food being fed. It is equally important to know how the feed is stored. Pellets should be stored in a cool, dry area away from direct sunlight. Rabbit or other rodent pellets are inappropriate as the sole diet for guinea pigs because they do not contain supplemental vitamin C. The best way to ensure vitamin C supplementation is to give vitamin C rich foods daily, such as kale or other dark leafy greens.

164 a They include intervertebral disc disease, discospondylitis, hypoglycemia, myelitis, diffuse muscular disease, diffuse skeletal disease, neoplasia, heart disease, anemia, hyperthermia, metabolic disease, GI disease and Aleutian disease.
b Collapse and hindlimb paresis can be a non-specific sign of disease in ferrets. Therefore, many diseases are considered when a ferret shows signs of paresis. The differential diagnosis is not limited to neuromuscular disease processes. Diagnosis of hindlimb paresis requires a thorough history, examination and possibly extensive testing. Perform hematology and serum biochemistry. A low RBC count (anemia) in ferrets is commonly caused by hyperestrogenism as seen in unspayed ferrets that remain in estrus or, rarely, ferrets with adrenal gland disease. Severe GI ulcers can also cause anemia. Renal disease and any chronic disease can result in anemia. An elevated WBC count may indicate a severe infection or lymphoma leading to weakness. A low blood glucose concentration is highly indicative of a pancreatic beta cell tumor. A fasting blood glucose (4–6 hours) concentration below 3.92 mmol/l (70 mg/dl) is good evidence that an insulinoma is present, especially if the ferret is above the age of three years. Ferrets with insulinoma commonly present with hindlimb paresis. Rule out other metabolic diseases based on the serum biochemistry results. Take radiographs to reveal the presence of cardiac disease and/or abdominal disease. Determine if disc disease is present by taking spinal radiographs or perform a myelogram.

This ferret was found to have a blood glucose concentration of 1.68 mmol/l (30 mg/dl) and an insulinoma was diagnosed. The ferret responded well initially to one dose of oral glucose and was then maintained on prednisone.

165 A two-year-old male ferret had progressive vision loss over the past two months. Both eyes appear the same (the left eye is shown (165)).
a What is your diagnosis?
b What etiologies have been suggested for this problem in ferrets?

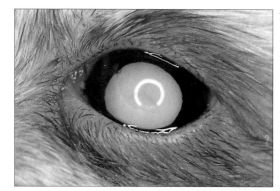

166 A ferret develops episodes of coughing and dyspnea. On physical examination, it is dehydrated and lethargic with pale mucous membranes. Thoracic radiographs reveal enlargement of the right ventricle and increased interstitial opacity throughout the lungs. Before further diagnostics are performed, the ferret dies. At necropsy, two helminths measuring 100 and 65 mm in length are recovered from the right ventricle (166).
a What are the worms?
b How was the ferret infected?
c What antemortem test can you carry out to confirm this infection?
d What treatment would you use for this infection?

167 A pet two-year-old male rabbit is shedding heavily and has patches of alopecia around the neck and ears. The lesions are approximately circular, have an epidermal collar and are mildly pruritic. The first lesions appeared two weeks ago and have increased in size.
a What is your diagnosis?
b How would you confirm this diagnosis?
c How would you treat the lesions?

165 a Bilateral mature cataracts are present. Normal ferret visual acuity is relatively poor (approximately 20/160), therefore if a ferret is adapted well to its environment before it becomes blind, the signs of vision loss may be subtle. Cataracts are a very common cause of blindness in ferrets with retinal degeneration the next most common cause. Cataracts typically began as multi-focal, punctate, white opacities in the axial region of the posterior lens cortex. These often progress to involve the anterior cortex and nucleus and, in some animals, the entire lens becomes opaque. In ferrets with long-standing cataracts, the lens may luxate into the anterior chamber and cause secondary glaucoma.
b Genetic and nutritional etiologies may be responsible for cataract development. Test breedings of affected animals suggest that there is a genetic component because offspring from X affected pairings may develop mature cataracts as early as 13–16 weeks of age. The exact mode of inheritance is unclear. A nutritional component is possibly involved in the genesis of cataracts in ferrets.

166 a *Dirofilaria immitis*, the dog heartworm.
b Ferrets are infected when mosquitoes bite them and deposit the infective third-stage larvae. The worms migrate into subcutaneous tissues and muscles. They reach the heart between 90 and 140 days postinfection.
c Microfilarial tests may be performed when heartworms are suspected, but only 20% of heartworm infections in ferrets are microfilaremic. Also, serum can be tested by one of the commercially available antigen-detection kits that detect heartworm antigen in the serum of dogs. While these tests are very specific and rarely give false-positives, ferrets may have too few worms in the heart to produce enough antigen to be detected by the tests. Therefore, false-negative results may occur when these tests are used for ferrets. Cardiomegaly may be visualized on chest radiographs and ultrasound may reveal the presence of worms.
d Infections may be treated with thiacetarsamide sodium (2.2 mg/kg q 12 hours for two days). It may be helpful to use either prednisolone (2.2 mg/kg PO q 24 hours for 3 months) or heparin (100 U SC q 24 hours for 21 days) along with thiacetarsamide to reduce the incidence of pulmonary emboli. To prevent infection effectively, ferrets can also be placed on a monthly dose of ivermectin (6 μg/kg PO).

167 a The most likely diagnosis is dermatomycosis caused by *Trichophyton mentagrophytes*. Differential diagnoses include hair loss due to molting and dermatitis due to ectoparasites or bacterial infection. *Microsporum* spp. has rarely been isolated from rabbits.
b By cytology and culture. Demonstrate the fungal agent on KOH preparations or fungal culture of skin scrapings and hairs from the infected area. *T. mentagrophytes* does not fluoresce under Wood's light illumination. Microscopically infected hairs contain large spores and mycelia. On DTM media, colonies are flat and yellow-orange on the underside.
c Treat localized lesions with topical 10% povidine or a topical anti-fungal cream or spray. Treat a generalized infection with griseofulvin (25 mg/kg PO q 24 hours for 14 days). Advise owners of the zoonotic potential of this disease. Treat other household pets as needed and clean the environment thoroughly.

168 Owners with rodent breeding colonies occasionally need to have the males castrated or vasectomized.
a How would you perform a castration in a rodent?
b How would you perform a vasectomy in a rodent?

169 A five-month-old intact female guinea pig has a hemorrhagic vaginal discharge. The owner thinks the guinea pig is five to six weeks pregnant. Only nine weeks ago the guinea pig delivered and raised three normal pups.
a What are your differential diagnoses?
b What do the radiographic findings reveal (169a, b)?
c How would you treat the guinea pig?

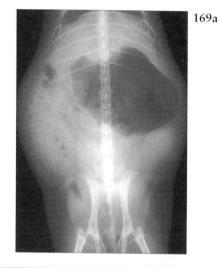

169a

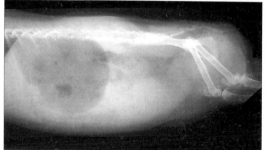

169b

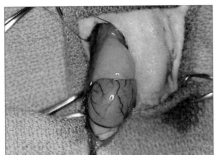

168a

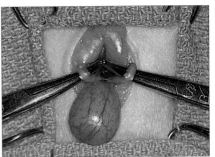

16

168 a The inguinal rings remain open throughout life in male mice, rats, guinea pigs and hamsters. Prevent inguinal hernias by using castration techniques that functionally close the inguinal rings. Rats, guinea pigs and hamsters have a large fat pad associated with each testis and spermatic cord. These structures make visualization and ligation of structures within the cord difficult. Use a standard closed castration technique in the smaller rodents. In larger rodents such as the guinea pig, one approach is to perform an open castration technique modified to close the inguinal ring. Place the anesthetized guinea pig in dorsal recumbency. Incise longitudinally over the scrotum and exteriorize the testis with the tunic intact (**168a**). Place a length of suture around the tunic and cord as close to the superficial inguinal ring as possible. Leave the suture untied as it will later identify and ligate the tunic. Incise the tunic distal to the untied suture, as shown. Clamp, ligate and remove the structures of the spermatic cord distal to the ligatures. Use synthetic absorbable suture. Allow the ligated stumps to slip back into the abdomen. Tie the suture that was previously placed proximally around the tunic. This functionally closes the superficial inguinal ring. Excise the distal tunic. Repeat the procedure on the other side. Leave scrotal incisions open or close with synthetic absorbable suture in a buried continuous intradermal pattern, or with skin staples or tissue adhesive.
b Perform vasectomy in rodents through bilateral scrotal incisions or via a suprapubic ventral midline incision. Make a small (2–3 cm) incision in the ventral midline just cranial to the pubis and take care to avoid injury to the bladder. Reflect the bladder caudoventrally, revealing the two vas deferens (**168b**). Double ligate each vas and remove a section between the ligatures. Close the linea alba with 4–0 or 5–0 synthetic absorbable suture. Close the skin with the same type of suture material in a buried continuous intradermal suture pattern, with Michel clips or skin staples.

169 a They include abortion, urinary tract infection, urinary tract calculi and genital tract infection.
b An indistinct radiopaque mass in the caudal abdomen in the region of the urinary bladder or uterus.
c Hospitalize the guinea pig for observation. Importantly, administer parental fluids and nutritional support and begin antibiotics that are safe for guinea pigs. Perform hematology and serum biochemistries. An ultrasound examination confirms the presence of soft-tissue changes in the uterus. The mass in the uterus has no skeletal features or active heart beats. The bladder appears normal. The diagnosis is intrauterine disease and the owner elects for an ovariohysterectomy.

170 This African hedgehog is displaying normal defensive behavior (**170a**). When threatened a hedgehog will draw its body tightly into a ball using the powerful orbicularis and panniculus carnosus muscles, thus making examination difficult.

a Describe several methods you could use to coax a hedgehog to uncurl.

b How would you administer medications?

170a

171

171 A pet rabbit has a severe conjunctivitis that has been present for several weeks (**171**). *Staphylococcus epidermidis* was found on a conjunctival swab culture. The antibiotic sensitivity test indicated that chlortetracycline was the drug of choice. After several weeks of topical treatment, there is no improvement in the ocular disease.

a Why might there be no response to therapy?

b How would you alter the treatment?

170b

170 a (1) Simply leave the pet undisturbed for several minutes on the table and it may uncurl and try to escape. (2) Hold the curled hedgehog in a gloved hand and stroke heavily from front to back over the caudal half of the body. (3) Hold the hedgehog's head downwards over a flat surface and it may unroll to reach for that surface. (4) Bounce the pet gently in cupped hands. (5) Place the hedgehog in a shallow (2.5 cm deep) container of warm water. This method should not be used if the pet is exhibiting any respiratory signs.

Once the awake hedgehog uncurls, scoop the pet up off the table in open hands and it will not usually curl up again. If all else fails, induce general anesthesia with either injectable drugs or inhalants. Isoflurane introduced with a mask or anesthetic chamber works well with rapid induction and recovery times.

Hedgehogs have spines that do not come out when handled. However, the spines are quite irritating when the patient holds them firmly erect in a defensive position. It is useful to protect your hands with latex examination gloves or lightweight leather gloves (170b).

b If the hedgehog is tractable or debilitated, you can grasp the loose skin behind the ears and suspend the patient over a table. This prevents the pet from rolling up and allows access to the mouth for administration of oral medications or feedings. In hedgehogs that are difficult to handle, use injectable drugs that can be given subcutaneously. Hedgehogs have an extensive subcutaneous space that contains a large amount of fat. Subcutaneous injections are given even when the pet is tightly curled into a ball. It is often not necessary to remove the animal from its cage. Mask oral medications with sweet flavors and put them in the patient's food; however, be aware that some pets will stop eating altogether if they sense there is a change in their diet.

171 a Conjunctivitis is not the only problem here. The discharge is more likely to be a sign of dacryocystitis. Consider the systemic health of the animal and its environment. Purulent conjunctivitis without bacterial involvement has been reported in animals exposed to environmental pollutants, such as dust from overhead hay racks. Culturing a conjunctival sac may not be entirely reliable. In this case, *S. epidermidis* may be part of the normal flora and is insignificant in the cause of the disease. To obtain more valuable information on possible bacterial involvment in cases of dacryocystitis, cannulate and flush the lacrimal duct with a sterile saline solution and collect the fluid as it exits the nares for microbiologic culture.

b Cannulate the nasolacrimal duct and flush with an appropriate antibiotic. Systemic antibiotics may also be necessary. Make necessary environmental changes. Treat any other underlying disease problems that may be affecting the overall health of the animal.

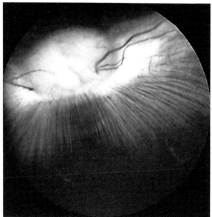

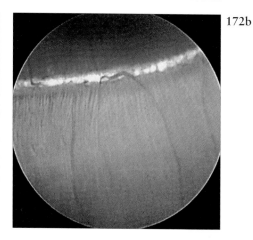

172 An examination of the fundus of a rabbit (172a) and a guinea pig (172b) reveals these images.
a What is your diagnosis for the rabbit?
b What is your diagnosis for the guinea pig?

173 An 18-month-old guinea pig is in respiratory distress. The pet is dyspneic with increased respiratory rate and effort. The oral mucous membranes appear mildly cyanotic and auscultation reveals muffled heart sounds with a respiratory wheeze and congestion. The signs have been present for four days, but are worse today. Recently, the owner changed the bedding materials from newspaper to cedar shavings. The guinea pig lives in a plastic cage with a wire top.
a Could the recent changes in husbandry have contributed to the signs?
b What are your differential diagnoses for dyspnea in a guinea pig?
c What diagnostic procedures and therapy are indicated?

172 a The fundus is normal for a rabbit. Lagomorphs possess a merangiotic retinal vascular pattern where vessels emanate from the optic nerve head and pass horizontally in association with the white medullary rays. The medullary rays represent myelinated nerve fibers. The optic nerve of the rabbit is far superior which results in a blind spot coincident with a view of the ground. This is a place from where few rabbit predators approach. The optic disc has a deep physiologic cup that should not be misdiagnosed as indicative of glaucoma.
b The fundus is normal for a guinea pig. Reportedly, the guinea pig has a paurangiotic retinal vascular pattern similar to the horse, but clinically it is considered to be anangiotic as retinal vessels are never observed. The orange stripes in the photograph represent choroidal vessels visualized through the avascular retina.

173 a Cedar shavings have been implicated as a cause of respiratory distress in guinea pigs due to the inhalation of high concentrations of various aromatic oils. These oils can be irritating to mucous membranes. Patients housed in environments with poor ventilation, such as aquariums or closed plastic cages, have an increased exposure.
b Ammonia toxicity secondary to poor hygiene and reduced ventilation, heat stress, bacterial or viral pneumonia, cardiac disease and neoplasia. Bacterial pneumonia caused by *Bordetella bronchiseptica* would be the primary differential in this case based on the provided history and signalment. This organism is commonly implicated in respiratory disease of guinea pigs, but it may be secondary to other stressors, such as environmental conditions, hypovitaminosis C or other disease conditions.
c Obtain a thoracic radiograph to determine the degree of pulmonary disease as well as any cardiac changes. Use extreme caution when restraining a dyspneic guinea pig for diagnostic procedures. Do not place these patients in ventral recumbency. Use sedation if necessary and always have oxygen available. If a significant amount of pleural effusion is detected, obtain a sample for cytology and microbiologic culture from a thoracic aspiration. Submit blood for hematology and serum biochemistry. Some animals suffering from toxicity due to aromatic oil inhalation will have elevations in serum liver enzymes. Perform an ultrasound of the thorax if needed, particularly in cases of cardiac disease or where an intrathoracic mass is suspected.
  Place the guinea pig in a well-ventilated cage with pelleted bedding made of a nontoxic material or newspaper. If an infectious disease is suspected, administer a safe broad-spectrum antibiotic, such as trimethoprim/sulfa, enrofloxacin or chloramphenicol. Use a diuretic, such as furosemide (2.2 mg/kg IV, SC, PO q 12 hours), to aid in thoracic fluid removal. A bronchodilating agent may also be used. Administer subcutaneous, intravenous or intraosseous fluids and gavage feedings of blenderized guinea pig pellets and fresh greens along with supplemental vitamin C (30–50 mg/kg PO q 12 hours), if needed. Provide oxygen if cyanosis is present. Handle the patient as little as possible and do not place it in a position that increases the pressure on the thorax.

174 A rat owner wishes to halt reproduction in her colony but does not want to spay her animals.
a What is one alternative to an ovariohysterectomy that can be used to curtail reproduction in female rodents?
b How would you perform this procedure?

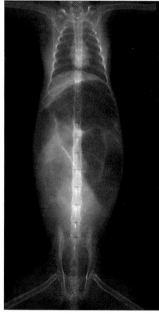

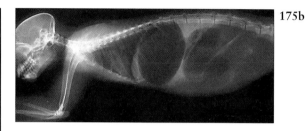

175 A five-year-old male cotton-top tamarin is found on the bottom of its cage. The monkey had a history of chronic intermittent diarrhea. The tamarin is now non-responsive, is in poor body condition, has abdominal distension and is emitting a foul-smelling, watery diarrhea. The monkey is hypothermic (temperature 36.7°C) and 10% dehydrated. The extreme abdominal distention prevents palpation of the viscera. The radiographs are shown (175a, b).
a What do these radiographs reveal?
b What are your differential diagnoses?
c What is your diagnosis and why?

174 a Perform ovariectomies in rodents as an alternative to ovariohysterectomy. Uterine disease is relatively infrequent in rodents and the voluminous abdominal contents in these animals make the approach to an ovariohysterectomy more difficult.

b Place the anesthetized animal in ventral recumbency. Clip and prepare the back for sterile surgery from the mid-thoracic to the caudal lumbar region. Keep the patient warm as hypothermia occurs from excessive clipping of hair or cooling by skin disinfectants. Several different surgical incisions are described for ovariectomies, each dependent upon the species and size of the rodent and surgeon preference. The ovaries are located in abdominal fat caudal to the kidneys. In very small rodents, such as mice, perform a single, small skin incision either longitudinally along the dorsal midline from the second to fifth lumbar vertebra or transversely at the level of the second lumbar vertebra. Use a transverse incision at the level of the flank glands in hamsters. With a single longitudinal incision, undermine the skin on each side to allow bilateral approaches through the flank muscles into the abdomen. In larger rats and guinea pigs, incise the skin caudal to the last rib and lateral and ventral to the paralumbar muscles (174). In mice, view the ovarian fat pads through the thin abdominal wall; this identifies where the abdomen is entered. In other species, use blunt dissection to penetrate the muscles of the flank lateral to the paralumbar muscles. Make a small incision into the peritoneum on each side. Large fat accumulations fixed to the paralumbar muscles are easily confused with periovarian fat, especially if the incisions are made too medially. Locate the ovaries and their associated, freely movable fat lateral to this paralumbar fat. Avoid disrupting ovarian tissue during surgery as detached portions could become functional. Grasp the periovarian fat to exteriorize the ovary such that the ovary itself is not handled. Some rodents have remnants of ovarian tissue associated with the uterine tube. Therefore, removing the entire uterine tube and cranial tip of each uterine horn prevents cycling. Ligate the ovarian vessels and locate the cranial tip of the uterine horn and ligate on each side. Between the ligatures, remove the ovaries, uterine tubes and cranial tips of each uterine horn. Close larger muscle incisions, such as those in the guinea pig, with synthetic absorbable sutures. Close skin incisions using either buried absorbable sutures, wound clips, staples or tissue adhesive.

175 a They reveal a loss of serosal margin detail throughout the abdomen. The entire GI tract is distended. The loss of serosal margin detail is likely the result of abdominal fluid or weight loss.

b For the greatly distended GI tract, either mechanical intestinal obstruction or ileus with gas distention. Causes of obstructive bowel disease include neoplasia, foreign body or stricture.

c Adenocarcinoma of the GI tract causing a functional obstruction is the most likely diagnosis. This neoplastic disease is common in middle-aged cotton-top tamarins and is often associated with chronic colitis. There can be multiple foci of adenocarcinoma in the cecum, colon and rectum but the most common sites are the colorectal junction and the ileum.

176 A guinea pig has difficulty walking. Physical examination reveals this lesion (176). The guinea pig is housed in a commercial wire cage that has a 1 × 2 cm mesh bottom with a piece of carpet placed in one corner. The diet consists of pellets and fresh vegetables. Water is available from a sipper tube.
a What is your diagnosis?
b How would you treat the condition?

177 An 18-month-old male hamster has bilateral alopecia and pruritus (177a). Skin scrapings reveal low numbers of *Demodex* spp. The hamster exhibits polyuria and polydipsia. Normal water intake for this tiny desert species is less than 10 ml/day. A metabolic cage is used to determine that the hamster is drinking up to 75 ml/day and urinating nearly the same amount.
a What differential diagnoses would you consider?
b What diagnostic tests should you perform?
c What is your diagnosis?

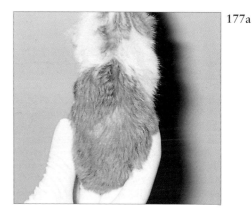

176 a Pododermatitis. Predisposing factors to this condition include rough cage flooring, unsanitary cage conditions, obesity, foot lacerations, hypovitaminosis C and any debilitating disease. *Staphylococcus aureus* is commonly isolated from these lesions but is usually secondary to these other factors. The feet appear swollen and discolored, and the foot pads are often cracked. If left untreated, the infection can lead to cellulitis, osteomyelitis and the eventual death of the animal.
b Manage uncomplicated cases (no cracks or fissures in the foot pads) with topical antibiotics, clean bedding and any additional husbandry improvements. Use protective ointments with a heavy petroleum base to keep these lesions moist and prevent cracks. Treat severe lesions more aggressively. Radiographs rule out osteomyelitis. In severe cases, administer systemic antibiotics, surgically debulk lesions, change bandages daily and soak affected limbs in a hypertonic or astringent solution.

177b

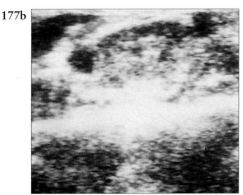

177 a Renal disease (including renal amyloidosis), demodectic mange, adrenal disease, hepatic disease, other endocrine disease or nutritional deficiencies. Demodectic mange causes hair loss and skin thickening. Adrenal disease can cause polyuria, polydipsia and hair loss due to hormonal effects. Nutritional deficiencies can cause a brittle, poor quality coat and patchy alopecia.
b Use serum biochemistry values, such as BUN and creatinine, in your diagnosis of renal disease. Manually express the bladder with great care or use cystocentesis to collect a urine sample and perform a complete urinalysis. Use multiple skin scrapings to demonstrate the relative number of demodectic mites. Examine the hair roots to determine if there is active growth or if they are in a telogen phase. Use radiography or ultrasonography to demonstrate abnormalities of the abdominal organs.
c Adrenal disease. Free catch urinalysis reveals a low specific gravity. Renal disease in this animal is less likely because serum biochemistry results indicate that the BUN and creatinine concentrations are normal. The skin is normal to thin with a significant number of broken or damaged hairs present. It was noted that most of the hairs were in the late telogen or resting phase. Demodex mites were found on skin scrapings but were present in low numbers. Scrapings of normal hamster skin often contain a small number of demodex mites. Ultrasonography confirmed the evidence of an enlarged adrenal gland and normal appearing kidneys (**177b**). The cystic adrenal gland is pictured here on the cranial medial pole of the kidney. Resolution of endocrine-associated clinical signs occurred when the adrenal gland was removed by flank laparotomy. A diagnosis of demodecosis in hamsters should always be investigated further. Infestation with large numbers of demodex mites is usually secondary to conditions that result in immunosuppression.

178 To improve airway access, it is important to be able to intubate rabbits. What three methods can you use for rabbit endotracheal intubation?

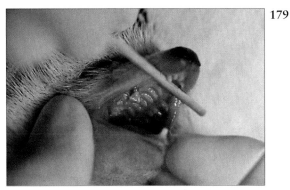

179

179 A two-year-old male African hedgehog is experiencing a steady decline in appetite (179). The diet consists of canned cat and dog food and some live insects.
a What condition is present in this hedgehog?
b How would you treat the condition?
c What are your recommendations to the owner to prevent it?

178a

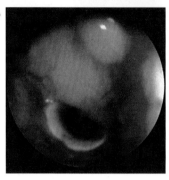
178b

178 Perform endotracheal intubation in rabbits using either a direct visualization, stylet-assisted or blind technique. For each method, position the rabbit in sternal recumbency with the neck extended. Pull the tongue through the diastema on one side so that it will not be cut or bruised by the teeth or laryngoscope blade (**178a**).

The direct visualization technique is suitable for use in larger rabbits. Introduce the lighted laryngoscope blade at the diastema and gently advance it over the base of the tongue until the larynx is brought into view (**178b**). Swab the larynx with a cotton-tipped applicator that has been soaked with a topical anesthetic if laryngospasm is a problem. Use a small amount of 1% lidocaine (approximately 0.25 ml). Benzocaine-containing spray has been associated with methemoglobinemia in rabbits and rats. Gently pass the endotracheal tube into the larynx and remove the laryngoscope. Use a semi-rigid atraumatic stylet to direct the tube for accurate placement.

Use the stylet-assisted technique in smaller rabbits because there is not enough room for simultaneous placement of the laryngoscope blade and the endotracheal tube. In these animals a flat laryngoscope blade (Miller) is easiest to introduce into the mouth to visualize the larynx. Pass a 3.5 or 5.0 Fr urinary catheter into the larynx, remove the laryngoscope and advance the endotracheal tube over the catheter into the airway. Quickly remove the catheter to prevent airway obstruction.

For the blind endotracheal intubation technique, position the animal with the neck extended as described above. Pass the endotracheal tube along the dorsal midline of the tongue toward the larynx. Listen for breath sounds and watch for condensation in the tube. Slip the tube into the larynx as the glottis opens during inspiration. Loss of breath sounds, loss of condensation and/or the presence of audible swallowing indicate that the tube is in the esophagus. The rabbit may respond to proper tube placement by moving or coughing. Confirm correct placement with the laryngoscope, by auscultation of the chest during lung inflation, or by testing for air passage through the tube.

179 a The receding gingiva, exposed root surfaces and bleeding gums indicate advanced periodontal disease.
b Perform a thorough physical examination along with appropriate diagnostic tests to rule out systemic causes of dental disease. Radiograph the skull to assess the health of the tooth roots and surrounding bone. Extract teeth that are loose. Scale and plane the remaining teeth and exposed roots. Polish the teeth and use appropriate antibiotic therapy. Institute an oral hygiene maintenance program.
c Dental disease in pet hedgehogs is most often caused by the accumulation of plaque. Prevent dental disease by providing a sound diet that includes firm foods, such as dry commercial cat food, hard-bodied insects and raw vegetables.

180 It is useful to anesthetize rodents for a variety of clinical procedures.
a What are some of the challenges of rodent anesthesia?
b What is the main difficulty you would encounter with the use of ketamine alone for anesthesia in rats, hamsters and mice?
c How would you use inhalant anesthesia in these species?

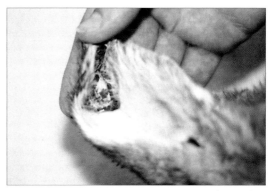

181

181 A three-year-old female rabbit is anorectic and constipated. Initially, it is treated with intravenous chloramphenicol, an intravenous-balanced electrolyte solution and multiple syringe feedings. The patient did well until four days later, when it develops a dry, leather-like ear tip (**181**).
a What are your differential diagnoses for the necrotic ear tip?
b How would you manage the condition?

180, 181: Answers

180 a The small size of these animals makes anesthesia challenging. The small size limits intramuscular and intravenous injection sites, causes difficulty in endotracheal intubation and increases the likelihood of hypothermia while anesthetized. Standard intraoperative monitoring is also challenging. Since species, strains and sexes may all differ in their response to the various anesthetics, it is difficult to extrapolate between different groups of rodents as to proper dosages of agents.

b Ketamine, used alone in rodents, causes such conditions as ataxia or cataleptoid anesthesia with retention of high muscle tone. The quality of anesthesia does not improve, even at very large doses in these animals. For this reason, combine ketamine with other agents to improve muscle relaxation and analgesia. Some agents that are combined with ketamine for rodent anesthesia include acepromazine, diazepam and xylazine. In some rodent species, the quality, depth and duration of anesthesia is quite unpredictable, even when ketamine is supplemented with other injectable medications. The regimens may induce excessive physiologic depression and prolong recovery at a dose needed to attain a surgical plane of anesthesia. Additionally, the use of multiple agents requires careful dose titration. Use a lower dose of each agent when in combination compared to the dose that is appropriate when the agent is used alone.

c Use isoflurane or halothane for rodent anesthesia. Isoflurane allows for a more rapid anesthetic induction, adjustment of anesthetic depth and recovery time, and is less of a cardiovascular depressant compared to halothane. Use inhalant anesthesia alone, after sedation or as a supplement to injectable anesthetics in rodents. Pretreatment with atropine (0.05 mg/kg SC) may be useful in some species (i.e. rats and guinea pigs) to decrease oral and respiratory secretions during induction with a volatile anesthetic.

Use a flow-through induction chamber in unpremedicated or lightly sedated rodents for anesthetic induction. Use a properly sized anesthesia mask and non-rebreathing circuit if a rodent is heavily sedated, lightly anesthetized or otherwise easily restrained to obtain a surgical plane of anesthesia. Chamber induction requires isoflurane concentrations of 3–5% and halothane concentrations of 3–4%. Use lower concentrations for mask induction. Continuously monitor animals during induction of anesthesia. Use the lowest concentration of the anesthetic agent that provides adequate anesthesia and muscle relaxation to maintain anesthesia. This ranges from 0.5–1.5% for halothane and 1.0–2.5% for isoflurane. The percentage depends upon the individual animal, other agents used and the method of administration.

181 a Phlebitis, trauma (especially in the long-eared breeds), disseminated intravascular coagulation, frostbite, fly or insect bites and cold-agglutinin disease. In this case, the most likely diagnosis is phlebitis with ischemic necrosis of the ear tip resulting from the intravenous catheter. This is a common sequela resulting from the use of aural intravenous catheters, especially in the short-eared breeds. Administration of irritating intravenous medications makes this condition more likely to occur.

b Surgically amputate the necrotic tissue. Place the patient on a broad-spectrum antibiotic for seven days. In addition, prevent contact with flies or other insects.

182 Venepuncture in the chinchilla is often necessary. What techniques would you use for collecting blood in the chinchilla?

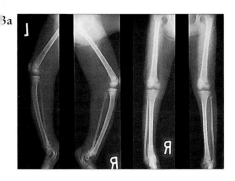

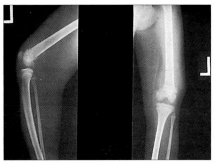

3a

183b

183 A juvenile male rhesus macaque has a two-day history of weakness and reluctance to use its thoracic limbs. Physical examination identifies instability at the ends of the long bones plus swelling and bruising of the arms. Diagnostic procedures include hematology and radiography (**183a, b**). Results of the hemogram are given in the table.

Values	Macaque	Normal values
PCV (%)	23	45.7–39.2
Hg (g/l)	82	12.5–14.4
MCV	62	72.4–77.6
WBC (× 10³/mm)	8.5	6.5–11.7
Neutrophils (%)	83	28–61
Eosinophils (%)	2	0.7–5.8
Lymphocytes (%)	13	34–67
Monocytes (%)	2	0–2.8

a What abnormalities are present on the hemogram?
b What are the abnormalities seen on the radiographs?
c What are the differential diagnoses?
d How would you confirm your diagnosis?
e How would you treat the macaque?

184 A 77 kg intact male potbellied boar has a three-day history of right forelimb lameness and lethargy. On palpation, there is pain, swelling and crepitus in the right elbow joint.
a What are your differential diagnoses?
b What diagnostic techniques would you use to evaluate the lameness?
c An intracondylar 'T' fracture of the distal humerus was diagnosed. How would you treat the condition?
d Is this a common problem?

182 The thick fur coat and relatively short front limbs renders venipuncture a challenge. In a laboratory setting, orbital sinus bleeding using a hematocrit tube is an efficient method for collecting small quantities of blood. This is effective but perform this only under general anesthesia. There is always the potential of damage to the orbital sinus if performed incorrectly. Do not use this technique on client owned animals.

In pet chinchillas, locate the small cephalic vein on the dorsum of the antebrachium. Do not shave the fur, rather, moisten the area with a small amount of alcohol. The lateral saphenous vein, which runs diagonally across the cranial aspect of the tarsus is perhaps the most convenient vein to access for both venipuncture and catheterization. If needed, collect large quantities of blood from the jugular veins. This is can be a challenge due to the chinchilla's short neck, thick fur and difficulties with restraint. Sedation or anesthesia may be necessary. Use a 25 gauge needle attached to a 1 or 3 ml syringe.

183 a The hemogram reveals anemia with microcytosis. The WBC is in the normal range but the neutrophilia and lymphocytopenia may be explained as a stress response.
b The radiographs reveal laterally displaced and/or collapsing physeal fractures in the distal femur, proximal tibia and distal tibia. A zone of rarefaction is present proximal to a line of increased density (known as the white line or scurvy line) across the metaphyseal region.
c Trauma or vitamin C deficiency are the two most likely differential diagnoses. The radiographic lesions are most typical of hypovitaminosis C.
d By an analysis of the food and diet. Do a vitamin C assay on serum or the buffy coat of centrifuged blood. Response to treatment is the most rapid and simplest way to support the diagnosis of hypovitaminosis C.
e Use vitamin C and iron supplementation, cage rest and analgesics as needed. Vitamin C contributes to the absorption, mobilization and use of iron. Hypovitaminosis C leads to microcytic, hypochromic anemia. Initially, give the monkey supportive care, such as fluids and force feedings, until it is stronger.

184 a They include trauma, fracture, luxation, neoplasia of the bone or joints, infection (septic arthritis) and degenerative joint disease.
b Use radiographs of the elbow and leg to determine the cause of the lameness. If no fractures or luxations are observed, perform a culture and cytology of fluid recovered from a joint aspiration.
c Treat this problem with surgical stabilization of the fracture. Reduce the fracture fragments to provide joint congruency and fracture repair. Use a lag screw technique through the condyles, followed by cross-pinning or plating of the humerus to provide good reduction and stabilization of this type of fracture.
d The bones of the forelegs are the most common place for fractures in potbellied pigs.

185 An adult rat is found dead in its cage unexpectedly. It had been treated topically by the owner for a minor dermatological problem. The stomach of the rat at necropsy is shown.
a What is the probable cause of death?
b How can the owner avoid this outcome in the future?

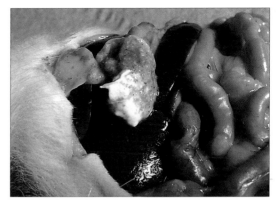

185

186 In rabbits the cecum is essential in the digestive process.
a What is unique about the cecum of the rabbit?
b What is the result of the fermentation of foodstuffs in the cecum?
c What dietary factors regulate cecotrophy?

187 A five-month-old male hamster has a firm swelling on the left side of the face. The owner noticed this two nights ago, and is now concerned because it appears to be getting larger. In addition, the pet is not eating or drinking normally.
a What anatomic structures may be affected to produce these signs?
b What are your differential diagnoses for this problem?
c Under what circumstances would this condition be considered normal?

185 a The cause of death is acute toxicity caused by ingestion of a dermatological preparation. The owner had applied excessive quantities of the preparation to the skin. Do not underestimate the toxic potential of topical preparations. Topical corticosteroids are absorbed systemically after ocular or auditory canal application. Antibiotic toxicity in rodents is reported from combined bacitracin-neomycin-polymyxin B dermatological preparations.

b Prevent this outcome by applying only a small quantity of medication to the skin lesion using a cotton-tipped applicator. Handle the animal for a brief period and then return it to its cage with a food treat to distract it. Use an Elizabethan collar to prevent ingestion in cases where large areas need to be treated topically.

186 a The rabbit has the largest cecum of any monogastric mammal. The cecum is a large blind sac. It contains about 10 times the volume of the stomach and 40% of the total volume of the GI tract. The serous membranes of the cecum are extremely thin and delicate compared to those of other sections of the bowel, and may tear easily when handled or sutured. It provides an anaerobic environment suitable for the fermentation of ingesta by a highly complex population of microorganisms.

b The result of fermentation in the cecum is called soft feces, night feces or ceco-tropes. Cecotropes, which are rich in nutrients such as fatty acids, amino acids and vitamins, are excreted separately from feces which are nutrient poor. By consuming cecotropes, rabbits ingest nutrients that would otherwise be lost. They are eaten directly from the rectum. The arrival of cecotropes at the anus triggers a neural response which results in licking the anal area and consumption of the cecotropes. Cecotropes are easily distinguished from feces. They are covered with a thin layer of mucus, are soft and moist, and tend to stick together and have a strong odor. Feces are hard, dry pellets.

c Cecotrope intake depends on the protein and fiber concentration of the diet. Cecotrophy is greater if the ration contains less protein and/or more fiber. Rabbits consume the total quantity of the produced cecotropes if the diet is energy deficient.

187 a Hamsters possess well-developed cheek pouches, which may appear as bulges on the sides of the face when filled or when diseased. Abnormalities of the jaw or teeth can also result in facial swelling.

b Normal food hoarding, food impaction, cheek pouch abscess or neoplasia, tooth root abscess, molar malocclusion, abscess or neoplasia of the mandible. It may be necessary to sedate the hamster in order to examine the area thoroughly.

c Hamsters normally hoard food by filling their cheek pouches and then hide their food in a secure location for later consumption. This behavior is amplified when a hamster is stressed or moved to a different environment. Other stressors that may result in this behavior include exposure to new cagemates, rearranging of cage furnishings or provision of excess food after a period of food deprivation.

The physical examination of this patient revealed food in the cheek pouch with no other abnormalities noted. The animal was recently moved to a new cage and then placed in a different area of the house. The resulting insecurity probably caused the hamster to hoard food more vigorously.

188 It is known that rabbits benefit from postoperative pain relief.
a What are some signs of pain and distress in the rabbit?
b What are some analgesics that you can use safely in the rabbit?

189 During the last two weeks, a six-year-old female guinea pig has shown aggression and sexual mounting behavior towards its female cagemate. The owner reports a weight gain and bilateral, multiloculated, round soft-tissue structures are palpated in the cranial abdomen. The CBC is normal except for a high percentage (6%) of the cell type seen in 189a.
a What is the name of the unusual cell type in 189a?
b What are these cells and what causes an increase in this type of cell?
c What does the abdominal ultrasound in 189b demonstrate?
d What are your differential diagnoses?
e How would you treat the condition?

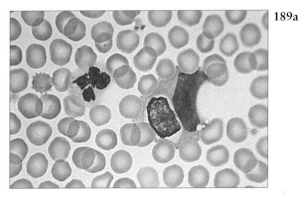

189a

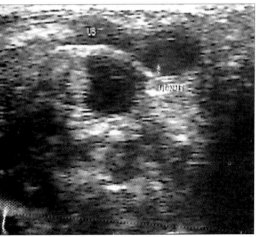

189b

190 Ketamine, in combination with medetomidine or xylazine, is a useful anesthetic for rodents and rabbits. What are three undesirable side-effects of this ketamine combination.

188 a There are both behavioral and physiologic signs of pain in rabbits. Severe, acute pain or fright is accompanied by a characteristic high-pitched squeal. Generally, any change in temperament or behavior from normal may be an indication of pain or distress. Specific behavioral conditions include inability to sleep, decreased activity, reluctance to move, timidity, depression, irritability, prolonged recumbency, lameness and abdominal splinting. Rabbits with abdominal pain may lie stretched out in the cage. A hunched posture and tooth grinding may also be seen in rabbits with abdominal discomfort. Two very important signs to watch for are decreased food and water intake. Physiologic findings that may indicate pain include hyperventilation, tachycardia, fluctuations in blood pressure and hyperglycemia.

b Although not approved for use in this species, a number of drugs have been reported efficacious or suggested for pain treatment in rabbits. These include opioid agonists, opioid agonist-antagonists and non-steroidal anti-inflammatory drugs. All have potential side-effects. Safe use requires an understanding of their pharmacologic properties. The opioid agonist, morphine (2–5 mg/kg SC, IM) produces good analgesia of 2–4 hours duration. Side-effects include ileus sedation and respiratory depression. Opioid agonist-antagonists (less side-effects than agonists) that are suggested for use in the rabbit include buprenorphine (0.03–0.05 mg/kg SC, IM), butorphanol (0.1–0.5 mg/kg SC, IM) and nalbuphine 1–2 mg/kg IM, IV). Non-steroidal inflammatory drugs, such as flunixin meglumine (0.3–1.0 mg/kg SC, IM), may be useful for treating mild to moderate musculoskeletal pain or pain due to inflammation.

189 a Foa-Kurloff cell or Kurloff body. The other cell is a neutrophil demonstrating the typical red granules found in rabbits and guinea pigs.

b Kurloff bodies are unique to hystricomorph rodents. They are eosinophilic intracytoplasmic inclusion bodies found in mononuclear leukocytes in the circulating blood, spleen, thymus, bone marrow and placenta. They are more numerous in females and tend to increase in number during pregnancy and after estrogen administration.

c A large septated cystic mass caudal to the right kidney. The mass measures about 2 cm in diameter. The body of the uterus is thickened and one of the horns is filled with fluid.

d They include cystic ovaries, cystic rete-ovarie or cystic neoplasia with secondary uterine disease, such as pyometra, mucometra or hydrometra. In this case, cystic rete-ovarie is the diagnosis.

e With an ovariohysterectomy. Cystic rete-ovarie have been associated with uterine tumors or hyperplasia of the endometrium. Since cystic rete-ovarie do not contribute to estrogen production, endometrial changes may be the explanation for the behavioral changes and the increase in Kurloff bodies.

190 Ketamine/medetomidine and ketamine/xylazine depresses respiratory and cardiovascular function. The result is hypoxia, hypercapnia and a fall in blood pressure. Thermoregulation is depressed and animals lose body heat. A specific side-effect of these combinations is hyperglycemia. Also, a marked diuresis results both from hyperglycemia and as a result of the inhibiting effects of xylazine or medetomidine on ADH secretion. Reversal of the xylazine or medetomidine with yohimbine or atipamezole respectively, often reverses many of these side-effects.

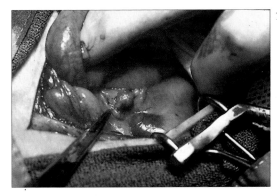

191

191 The abdomen of a six-year-old ferret during an exploratory laparotomy (191). A tumor of the right adrenal gland is present.
a What are some clinical signs that are seen in ferrets with adrenal gland tumors?
b Describe the surgery to remove the adrenal gland tumor. Why are right-sided adrenal gland tumors more challenging to remove than left-sided tumors?

192 In late autumn, a juvenile European hedgehog weighing 300 g exhibits marked dyspnea with open mouth breathing. There is some foam visible at the lip commissures and the oral mucous membranes are cyanotic.
a What life-saving treatment would you initiate?
b What is your diagnosis ?
c What further treatment would you institute?
d What long-term management is needed?

191 a Signs of adrenal gland disease in ferrets include bilaterally symmetrical alopecia. Pruritus, sexual behavior and an increase in body odor are also present in some ferrets with adrenal gland tumors. A specific sign of this type of tumor in the female ferret is vulvar enlargement. Male ferrets may present with dysuria, a caudal abdominal mass and urinary tract blockage due to cystic hyperplasia of the glandular prostatic urethra.
b Perform an exploratory laparotomy. Visualize and palpate both adrenal glands. The normal adrenal glands are 5–8 mm long, 4–5 mm wide and 3–4 mm thick. The left adrenal gland is located craniomedial to the cranial pole of the kidney and is usually surrounded in fat. The main blood supply to the left adrenal gland is the phrenicoabdominal vein which courses ventrolaterally across it. When removing the left adrenal, ligate this vein using hemostatic clips as the tissues are friable and may not hold sutures well. Clamp the phrenicoabdominal vein medial and lateral to the left adrenal gland. Gently dissect the gland away from the retroperitoneal tissues. Control capillary bleeding with a hemostatic sponge. The right adrenal gland is more difficult to remove since it is partially under the liver and is adherent to the caudal vena cava, craniomedial to the cranial pole of the right kidney. Attempts to 'tease' the adrenal gland off the vena cava usually result in tearing of the vessel. Remove the adrenal gland by placing hemostatic clips between it and the lateral wall of the vena cava and carefully cut away the gland. Alternatively, debulk the adrenal gland by incising its lateral surface and scooping it out of the subcapular contents. Be aware that this method can result in severe bleeding and death if venous sinuses are present. If both adrenal glands are abnormal, remove the larger of the two and do a subtotal adrenalectomy of the other gland. Biopsy the liver and any abnormal abdominal structures. Examine the pancreas for evidence of insulinoma. Close the abdomen in a routine manner. Maintain the ferret on intravenous fluids postsurgically until stable. Administer postoperative steroids if more than one adrenal gland is removed or if the ferret is lethargic with no obvious cause.

192 a Use an oxygen cage or chamber to minimize stress when treating cyanosis. Administer furosemide (2.5–5.0 mg/kg SC). Administer an injectable antibiotic such as amoxicillin to cover the potential for bacterial disease.
b Lungworm infestation. In hedgehogs this is usually a mixed infestation of *Crenosoma striatum* and *Capillaria aerophilia*. A thorough history should establish if paraquat poisoning, the most likely other diagnosis, is possible. The finding of *Capillaria* eggs in the feces does not rule out paraquat poisoning because intestinal *Capillaria* can produce such findings. Most commercial laboratories that deal with bovine or ovine species should be able to analyze hedgehog feces. In heavy infestations, direct examinations of a wet preparation of feces under the microscope may reveal *Crenosoma* larvae.
c Often there is a respiratory infection superimposed on the lungworm infection. *Bordetella bronchiseptica* is frequently isolated at postmortem. To maximize the chances of successful treatment, continue the course of antibiotic therapy Once the patient is stable, administer ivermectin (0.2–0.4 mg/kg SC). Alternatively, give levamisole hydrochloride (10–20 mg/kg SC).
d A hedgehog weighing 300 g is too small to hibernate so it will need to be over-wintered in a warm environment and provided food on a daily basis. To prevent a recurrence of the respiratory disease, repeat the ivermectin or levamisole injection at monthly intervals.

193 Disease prophylaxis is accomplished in pet potbellied pigs via vaccination. What vaccinations would you recommend for the pet potbellied pig?

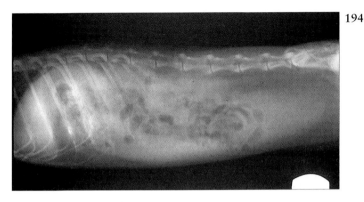

194

194 A four-year-old male castrated ferret was examined because of progressive stranguria and urinary incontinence of two weeks duration. The ferret also had a two-month history of dorsal pruritus. Physical examination revealed a large, painful bladder that is difficult to express. A radiograph is taken (**194**).
a What does the radiograph reveal?
b What are your differential diagnoses for stranguria associated with a large bladder in ferrets?
c What tests would you carry out to diagnose the cause?
d How would you treat the condition?

193 Although many of the porcine diseases share similar names as diseases in domestic pets, never use dog and cat vaccines in pet potbellied pigs. For example, *Bordetella bronchiseptica*, one of the bacterial pathogens responsible for kennel cough in dogs, is a different strain from that which causes atrophic rhinitis in pigs. There are far more diseases in pigs that have vaccines available than there are for dogs and cats. However, not all of these diseases need be protected against in the private or home environment.

Recommendations range from no vaccinations to multiple inoculations. Adapt the vaccination protocol to the specific pet and its living conditions. Rabies virus vaccination is generally not considered necessary for pet pigs. Currently, there is no rabies virus vaccination approved for pigs in the USA. For single pig or isolated households where there is no exposure of pet pigs to outside pigs, leptospirosis and erysipelas are considered to be the minimal prophylaxis required. Initially vaccinate at weaning and boost annually.

Breeder pigs may benefit from vaccinations against porcine parvo virus, as well as the antidiarrheals, such as collibacillosis, rotavirus, *Clostridium perfringins*, TGE and salmonellosis. Vaccinate the breeding sow six weeks prior to parturition.

In commercial or breeder colonies, respiratory disease is a significant problem. Protect pigs by vaccinating against *B. bronchiseptica*, *Pasteurella multocida* types A and D, *Mycoplasma hyopneumoniae* and *Actinobacillus pleuropneumoniae*. Start these vaccinations when the piglet is a week old.

Many companies produce polyvalent products that contain several of the above vaccines in combination. When using any polyvalent vaccine, always follow label instructions.

194 a The presence of a large bladder.
b Urinary tract calculi, urinary tract infection, neoplasia and constriction of the urethral passageway due to diseased prostatic tissue along the urethra caused by infection, cysts, hyperplasia or cancer. Urethral stricture in male ferrets with adrenal gland disease is caused by enlargement of the prostatic tissue surrounding the urethra.
c Collect urine by cystocentesis for urinalysis and bacterial culture and sensitivity testing. Determine the plasma concentration of adrenal hormones to aid in the diagnosis of adrenal gland disease. Use abdominal ultrasound to detect the cause of the enlarged bladder and to detect any adrenal gland enlargement. In this case, abdominal ultrasound revealed a large right adrenal gland and a swelling along the urethra just distal to the bladder. Determination of plasma concentration of adrenal hormones revealed high androstenedione and estrogen which is consistent with adrenal gland disease in ferrets.
d Initial treatment includes administration of antibiotics and placement of a urinary catheter and collection system under anesthesia. Once stable, perform an exploratory laparotomy to remove the enlarged adrenal gland. In this case, the periurethral tissue was bilaterally swollen just distal to the bladder. Incisional biopsy of the mass revealed cystic prostatic tissue. No attempt was made to remove this tissue, because without the presence of infection the swelling dissipates when the adrenal gland is removed. The ferret recovered uneventfully, and the urinary catheter system was removed 12 hours after surgery. The prognosis is guarded because prostatic enlargement may recur.

195 An adult female macaque develops an ulcerative lesion on her upper lip and bilateral blepharoconjunctivitis. She also has a week-old infant with conjunctivitis and periorbital lesions consisting of small, flattened blisters.
a What is the most likely etiology of the lesions?
b What are the most common presenting signs of this agent?
c What is the zoonotic significance of this agent?

196

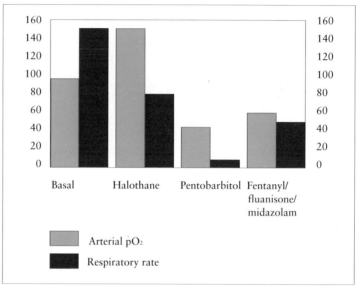

196 This graph shows the changes in arterial oxygen tension and respiratory rate following production of surgical anesthesia in rabbits with halothane, pentobarbitol or fentanyl/fluanisone/midazolam. What do these changes illustrate?

197 An adult rabbit has discolored urine. A complete physical examination and appropriate clinical pathology is performed and the diagnosis is uncomplicated bacterial cystitis. A bacterial agent is isolated on urine culture that is sensitive to trimethoprim/sulfadiazine. There is no improvement despite two weeks of treatment using an oral trimethoprim/sulfa preparation once daily. What factors may be causing the treatment failure?

195 a The most likely etiologic agent is *Herpesvirus simiae* (B-virus).
b *H. simiae* is uncommonly associated with active disease. It is usually an asymptomatic infection in macaques. The most commonly seen signs are localized, mild lesions consisting of small oral or mucocutaneous vesicles or ulcers. Also observed in adult monkeys are disseminated infections or CNS lesions. *H. simiae* is activated during pregnancy as evidenced by recovery of virus from genital swabs. *H. simiae* disease in the infant macaque consists of ulcerative conjunctivitis, vesiculo-ulcerative lesions on the face and head, and disseminated lesions in the liver and adrenal medulla. Histopathologically, this is characterized by necrotic foci with intranuclear inclusions and meningoencephalomyelitis throughout the brain and spinal cord.
c *H. simiae* causes a fatal encephalomyelitis in people. The virus is transmitted to humans during contact with macaque saliva, genital secretions, blood or mucus. Virus shedding is intermittent and no clinical signs may be visible at the time of shedding. But monkeys with this disease are considered infected for life. The most important aspects of the guidelines for prevention of this disease in people are to wear protective clothing, and immediately and vigorously cleanse monkey bites or scratch wounds.

196 The rabbit anesthetized with halothane will almost certainly be receiving 60–100% oxygen as the carrier gas, and so naturally develops a marked elevation of pO_2, despite a reduction in respiratory rate. In contrast, rabbits receiving the injectable anesthetics show both a fall in respiratory rate and a substantial fall in pO_2. This is due to failure to supplement the animals with oxygen. A common misconception is that following use of an injected anesthetic, no special measures need be taken to supply oxygen. Providing oxygen by face mask can be life-saving if the animal has pre-existing respiratory disease. Although hypercapnia is not prevented, prevention of hypoxia is of significant benefit.

197 The most common problem is that an inadequate dose and dosage frequency are used. Smaller species may require a higher dose and more than once daily treatment. Another problem is client compliance with the treatment regimen. It can be difficult to administer oral medication to a pet, particularly if it is unpalatable. In addition, some owners' personal schedules do not allow frequent dosing of medication.

It is necessary to use the appropriate antibiotic as shown on sensitivity testing. In this case, trimethoprim/sulfadiazine was being used. However, sulfadiazine has been shown to have a half-life of 40 minutes in some rabbits, and it has an inactive metabolite. This means that for most of the day the only active antibiotic is trimethoprim. This illustrates that although metabolic scaling can give a dosing regime, major variations in the metabolism of an individual drug in a particular species cannot always be predicted. Monitor all animals regularly during the treatment period to determine efficacy of treatment and potential dangerous side-effects. Rabbits and some rodents can develop fatal antibiotic-induced enterocolitis.

198 The owner of a five-year-old guinea pig asks that a CBC and serum biochemistry profile be performed to make sure that the pet is in good health.
a Why is it not appropriate to scruff the guinea pig, as is done with hamsters and ferrets, to aid in examination?
b Which veins are accessible for venipuncture ?
c How would you collect the blood?

199

199 A gerbil has a large bleeding mass on its ventrum (199). This non-pruritic lesion has been slowly increasing in size for three weeks. The gerbil is housed in a plastic cage with wood shavings for bedding.
a What is your diagnosis?
b What is your prognosis and how would you treat the condition?
c Masses such as these arise from what structure?
d Are there any other dermatological conditions associated with this structure in the gerbil?

198

198 a The guinea pig does not have excess skin over the nape to allow scruffing. The skin is tightly attached to the underlying tissues, particularly the subcutaneous fat pad over the nape of the neck. Guinea pigs appear to be more sensitive than other species and exhibit a painful response to scruffing. In addition, guinea pigs become distressed when they are placed in dorsal recumbency. When examining the ventrum of the awake patient, always keep the head elevated. Some guinea pigs will become so stressed that they will become unconscious during an examination or when a diagnostic procedure is attempted. Always have oxygen available for these situations and advise the owner of potential underlying problems, such as cardiac disease, pulmonary disease or anemia.

b The cephalic, lateral saphenous, femoral and jugular veins can be accessed in the guinea pig. Depending on the condition and co-operation of the patient, it may be necessary to use sedation to obtain the sample.

c The procedure for bleeding from the cephalic, femoral or lateral saphenous vein is very similar to that of other species. The photograph (198) shows the lateral saphenous vein. Hold off the vein manually or with a tourniquet and use a 25–27 gauge needle and insulin syringe to obtain the sample. Once blood is observed in the hub of the needle, release the pressure on the vein to allow more blood to enter the area. Use gentle traction on the plunger of the syringe to avoid collapsing the vein. An alternative method is to use the needle without a syringe attached and allow the blood to flow directly from the hub into a collecting tube.

Approach the jugular vein with the sedated patient in dorsal recumbency. It is difficult to palpate the vein due to the thick and muscular neck of the guinea pig. The jugular vein lies in a furrow running from the thoracic inlet to the base of the ear. Hold off the vein at the level of the thoracic inlet and use a 22–25 gauge needle attached to a syringe to draw the blood sample.

199 a This mass is most likely a sebaceous gland adenoma. Sebaceous gland adenocarcinoma is also a possibility but is found less frequently. Skin neoplasms usually occur in aged gerbils.

b Sebaceous gland adenoma is a benign neoplasm and carries a good prognosis if resected fully. The prognosis is guarded to poor for sebaceous gland adenocarcinoma because metastasis to distant sites and local recurrence is common. Surgical removal is the only treatment option.

c These masses often arise from the ventral scent gland of the gerbil.

d The ventral scent gland can become inflamed when rubbed on abrasive surfaces. In addition, apparent impaction of the scent gland leading to self-trauma has been reported.

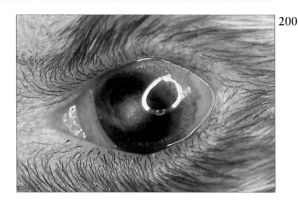

200

200 The owner of a pet shop has a six-month-old female ferret with abnormal eyes. Both eyes appear similar and have been like this since they obtained the animal at 16 weeks of age. The right eye is shown (200).
a What is abnormal about the eye?
b What is your differential diagnosis?
c What causes this defect?

201 It is necessary to perform thorough physical examinations on pet rabbits.
a How would you restrain a rabbit during a physical examination?
b How would you administer oral and parenteral medications in the rabbit?

200 a The eye is grossly smaller than normal (microphthalmic) and affected by a bizarrely vacuolated, mature cataract. On close examination, the iris is thicker and more lacy in appearance than normal. Histopathology demonstrates that the entire eye is small and that iridal development is arrested. The ciliary body originates from the posterior surface of the iris rather than from its normal position posterior to the iris root and the lens is comprised of large vacuoles and has extruded through the posterior capsule into the vitreous. The retina is dysplastic in areas where traction was applied to its surface (such as the peripheral retina near the ciliary body and the optic nerve head).
b A differential diagnosis includes long-standing, lens-induced anterior uveitis and subsequent phthisis bulbi. The animal's young age, the presence of bizarre vacuoles in the lens and the lack of inflammation on histopathology indicates that this is a congenital problem.
c Test breedings over four generations demonstrate this to be an autosomal dominant trait. Other organ systems are grossly normal but it is unclear whether homozygotes are viable. The degree of microphthalmia is variable but affected animals have bilaterally abnormally small eyes. Retinal degeneration also accompanies the microphthalmia because the retina develops normally until several weeks of age after birth and then degenerates. To eliminate the disease, do not breed affected animals.

201

201 a A towel wrap is a simple method of restraint (**201**). Place the rabbit on the edge of an appropriately sized towel and tightly fold up one side, then the back and finally the second side. Make sure the feet are tucked inside the wrap. To keep the rabbit quiet during the wrapping process, place a hand over the rabbit's eyes. The rabbit can then be restrained for examination, medication, and minor procedures, such as tear duct flushes, ear cleaning and mouth examinations. A cat restraint bag can be used in a similar manner for larger breeds of rabbits.
b Some rabbits can be medicated orally by hiding crushed tablets or liquids in sweet-tasting substances such as apple sauce, fruit juice sweetened jam, molasses and fruit or vegetable juice. Use the towel wrap to restrain the patient for direct administration of oral medications or feedings. Some rabbits will take whole tablets if they are pushed through to the back of the mouth.

Give subcutaneous injections anywhere along the dorsum. Intramuscular injections are given in heavily muscled areas, such as the quadriceps, semimembranosus, semitendinosus or lumbar muscles. Administer intravenous injections in the cephalic or lateral saphenous veins. Do not routinely use the auricular artery or vein for injections because of the potential for dermal slough.

202 An adult female Virginia opossum is hit by a car and sustains a simple transverse midshaft tibial fracture. She is in good body condition and has normal mucous membrane color, but is depressed and has a pouch full of very small babies.
a Should the opossum be handled?
b Can anesthesia be used safely?
c How will you repair the fracture?

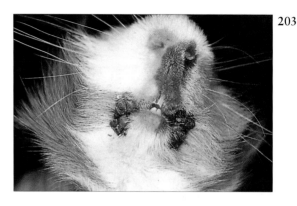

203

203 An adult guinea pig developed these lesions gradually over a period of two weeks (203).
a What is your diagnosis?
b What organisms might be involved in these lesions?
c How would you treat this condition?
d What management changes would you make?

202

202 a Use minimal handling until the patient is stabilized. The patient is not only a wild animal in pain, but it is also caring for young. Opossums may reject their young if subjected to excessive stress. Obtain a weight and place the patient in a warmed cage or incubator well padded with towels. Cover the cage to allow for privacy. Administer subcutaneous fluids and corticosteroids to treat shock. Give an injectable antibiotic if there is evidence of infection, such as in the case of a open fracture. Use analgesics with caution in nursing mothers. If the patient is eating and its behavior is normal, analgesics are probably not necessary. Obtain diagnostic samples only if the procedures can be tolerated without excessive restraint. Use general anesthesia to minimize stress on fractious or aggressive patients to perform examinations and diagnostic testing.

b Anesthesia can be used safely on the stabilized opossum. Isoflurane is a safe and effective anesthetic agent. Induce anesthesia with a mask or in an anesthetic chamber and maintain with a mask or endotracheal tube. Other anesthetics used in this species include halothane, metofane and ketamine and diazepam. This opossum was anesthetized with isoflurane the day after admission into the hospital to allow for a complete physical examination and radiography. She did not reject her young, although this is a possibility that has to be weighed when considering anesthesia in nursing mothers of this species.

c Use either internal or external fixation devices based on orthopedic principles. If a nursing mother undergoes a long surgical procedure there is the potential that she may reject her young post-surgically. In this animal, a lateral splint was applied under isoflurane anesthesia (202). Recovery was uneventful. The patient readily accepted a diet of canned dog food and canine milk substitute. She tolerated once a day handling and cleaning and continued to nurse her young. The splint was removed at four weeks when the fracture had healed. One week later she was returned to the wild with her ten young hanging on her back.

203 a Cheilitis (inflammation of the lips). Lesions are usually limited to the commisures of the lips.

b The organism most commonly isolated from these lesions is *Staphylococcus aureus*. However, the pathogenesis of this condition is often multifactorial and a pox virus has recently been isolated from similar lesions in guinea pigs in the UK. Predisposing factors in this disease include abrasive foodstuffs or bedding (such as wood chip bedding and poor quality wheat straw), water bottles with damaged nozzles, dental disease and hypovitaminosis C. Feeding good quality grass hay is not a risk factor.

c Gently debride the affected areas, culture and then clean with a mild antiseptic solution. Use topical and systemic antibiotics appropriate for the guinea pig. Treat any dental disease present.

d Remove predisposing factors. Add half to one cup of dark leafy greens, such as kale, dandelion greens, mustard greens, beet tops or collard greens, per guinea pig to provide a natural source of vitamin C. Use additional vitamin C supplementation orally in severely affected cases to improve healing.

204 Improper husbandry can contribute to many clinical problems in the rabbit, such as pododermatitis, respiratory disease and enteritis. What is a suitable environment for a rabbit?

205

205 This wild European hedgehog is picked up out of the backyard by a concerned homeowner (205). It has a number of non-pruritic, scaling lesions most prominently on the head. The lesions do not fluoresce under a Wood's lamp.
a What is your diagnosis?
b How would you confirm this diagnosis?
c What treatment would you recommend for this condition?
d Is there a zoonotic potential for this disease?

204

204 The optimal temperature range for the rabbit is 16–21°C. They may be housed indoors or outdoors. If housed outdoors, protect the rabbit from the elements, extremes in temperature, and from predators. An appropriate cage size should average about 0.30 m² per kilogram of body weight. The best cage material is galvanized wire or stainless steel. Cages made of wood are destroyed by chewing and are difficult to disinfect. Provide flooring that can be kept clean and dry. Wire mesh floors with 1 × 2.5 cm openings, allow feces and urine to fall through to a drop pan (204). However, if the spacing of the wire is too wide, foot and toe injuries can occur. Provide a cloth pad, newspaper or a box filled with bedding in one area of the cage to give the pet a resting place off the wire. Solid-floored cages can be used but are more difficult to keep clean. Change bedding such as straw, hay or pelleted products at least twice a week to prevent feces and urine accumulation.

Rabbits tend to eliminate in one corner of the cage. They can be trained to use a litter box by placing the box containing an absorbent bedding in preferred areas. Do not use clay cat litter in the box because some rabbits will ingest this material and develop a fatal gastric or intestinal impaction. Allow the rabbit a minimum of one to two hours daily to exercise outside its cage in a supervised area. It is necessary to 'rabbit-proof' areas to which the rabbit has access to prevent the destruction of electrical cords or furniture. Rabbits enjoy a variety of toys as a means of satisfying their need to chew and to combat boredom. Acceptable toys include untreated straw mats or baskets, cardboard boxes or tubes, paper bags, untreated wood scraps, hard plastic baby toys, hard rubber or metal balls containing bells, jar lids and cardboard or plastic boxes filled with hay or straw used for digging.

205 a This is a case of ringworm caused by *Trichophyton erinacei*.
b Culture a scab from the lesion on dermatophyte test medium. A heavy growth of *T. erinacei* is evidence of clinical disease. Other differential diagnoses include ectoparasites and bacterial skin infection. About 25% of wild hedgehogs of this species are asymptomatic carriers of *T. erinacei*.
c Administer griseofulvin (15–50 mg/kg q 24 hours) for several weeks. For wild hedgehogs accustomed to feeding from a bowl, advise the homeowner to give the daily dose of medication in a little food. A chocolate flavored pediatric elixir is particularly well accepted by hedgehogs. For captive hedgehogs, spray a solution of enilconazole (Imaverol, Janssen) on the lesions to assist in clearing up the infection and reduce the infectivity of the hedgehog to others. Use the spray once every 3–4 days for 3–4 applications.. This product can also be used to treat the environment.
d Yes. *T. erinacei* can be transmitted to dogs and humans. In dogs, the lesions appear on the lips or muzzle either from investigating hedgehogs found while walking or by eating left-over food from a bowl at which an infected hedgehog has been feeding. The lesions on humans can be intensely pruritic, but may not resemble classic ringworm and therefore not be recognized by a medical practitioner. People working with wild hedgehogs can contract dermatophytosis without having encountered an animal with obvious lesions due to the presence of asymptomatic carriers.

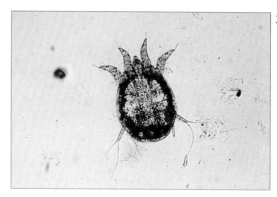

206

206 A rabbit living in an outdoor hutch is constantly shaking its head and scratching its ears. There is crusting in the external ear canal. On physical examination, the skin under the crusts is erythematous and moist. The organism shown is found on a mineral oil scraping of the debris (206).
a What is the organism and is it responsible for the head shaking and ear scratching?
b How would you treat this condition?

207 A four-year-old ferret has increasing episodes of lethargy accompanied by hypersalivation. On physical examination, no abnormalities are noted. The CBC and serum biochemistry panel are normal except for a fasting blood glucose of 1.96 mmol/l (35 mg/dl). Radiographs are normal.
a What is your diagnosis?
b What other tests would you perform?

208 An African hedgehog requires anesthesia for removal of a mass on the foot. Physical examination reveals an otherwise healthy animal. By what methods can anesthesia be achieved in the hedgehog?

179

206 a The rabbit ear mite, *Psoroptes cuniculi*, which causes inflammation and crusting in the external ear canal. This condition is sometimes called 'canker'. Identify *Psoroptes* mites by their long legs with suckers on a long, jointed stalk on pairs 1, 2, and 4 in the female and 1, 2, and 3 in the male.
b With ivermectin (0.20–0.40 mg/kg SC) or placed directly into the ear canal over the lesion at 10–14 day intervals for three doses. The ears do not need to be cleaned because the ulcerations will heal and the crusts fall out within 10–14 days of the initial treatment. *P. cuniculi* causes a less severe form of otic acariasis in horses and goats.

207 a Insulinoma (pancreatic beta cell tumor) is the most likely diagnosis. This is a common disease in ferrets over the age of three years. Signs of this disease are related to hypoglycemia and include intermittent lethargy, hypersalivation, weakness and pawing at the mouth. Pawing at the mouth and hypersalivation may be indications of nausea.
b Most ferrets with this disease have normal hematologic and serum biochemistry parameters except for a lowered fasting blood glucose concentration. Ferrets should be fasted a minimum of four hours to a maximum of six hours. Normal fasting glucose concentration range is approximately 5.0–6.7 mmol/l (90–120 mg/dl). It may be necessary to take more than one fasting sample to demonstrate a lowered blood glucose concentration. A plasma insulin concentration may help to identify ferrets with this disease. To interpret the results correctly, each laboratory must validate their own test for use in ferrets. An inappropriately elevated plasma insulin concentration in association with a low blood glucose concentration is indicative for insulinoma. A normal blood insulin concentration does not mean that an insulinoma is not present. A normal insulin concentration can be seen with a very low blood glucose concentration in ferrets. Abdominal ultrasonography infrequently detects pancreatic tumors. The definitive method to diagnose this disease is with exploratory surgery.

208

208 Isoflurane provides the easiest and most effective means of achieving anesthesia in many situations. Anesthetize hedgehogs using an induction chamber with 100% oxygen at an isoflurane concentration of 3–5% (**208**). Once induced, maintain the hedgehog with either an endotracheal tube or face mask at an isoflurane concentration of 0.5–3.0%. Although not generally needed, atropine (0.02–0.04 mg/kg IM, SC) can be used as a premedication. Injectable anesthetic agents produce more prolonged and sometimes more difficult recovery times than inhalant anesthesia. One choice is ketamine (5–20 mg/kg IM) either alone or with diazepam (0.5–2.0 mg/kg IM). Another combination is tiletamine/zolazepam (1–5 mg/kg of the combined drug IM). There is a large amount of subcutaneous fat over the dorsum. Injection into this area can result in unpredictable absorption

209 A four-year-old female guinea pig has a history of five days of stranguria, hematuria and lethargy. She is on a diet of guinea pig pellets, vitamin C in the water and alfalfa cubes. This radiograph of the abdomen revealed a 1 × 2 cm mineralized density in the urinary bladder (**209**). Urinalysis via cystocentesis is obtained (below).

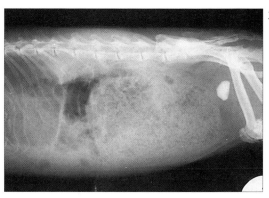

Urinalysis

Color	Red
Appearance	Opaque
Sp. gravity	1.020
pH	9.0
Protein	3+
Glucose	Negative
Ketone	Negative
Bilirubin	Negative
Hemoglobin	4+

Urine sediment

Epithelial cells	Transitional
Casts	None
Crystals	Amorphous and calcium carbonate (moderate amount)
RBC	50–60/HPF
WBC	10–20/HPF
Bacteria	Few

a What is your diagnosis?
b How would you carry out the surgical treatment of this problem?
c What are the densities most likely composed of?
d What is the prognosis?

210 Two young wild rabbits are found in a suburban backyard (one of them is shown (**210**)). They are normal in appearance with no outward signs of trauma and their eyes are still closed.
a What advice can you give to the captor?
b What formula can be used to feed the animals and how frequently should they be fed?
c At what age will they open their eyes?

209 a Calculi in the urinary bladder and right distal ureter.
b Exploratory laparotomy is the treatment of choice for this problem. Evaluate renal function prior to surgery. Palpate the ureters for calculi. Attempts to dislodge a calculus into the bladder are usually not successful. A ureterotomy may be done to remove a calculus but postoperatively strictures can occur. If the ureteral calculus has caused unilateral hydronephrosis, remove the kidney and ureter. Perform a cystotomy to remove the calculus from the urinary bladder. Exteriorize the bladder and pack it off with moistened gauze sponges. Make an incision in the apex of the bladder and remove the calculi. Culture the bladder wall. Lavage the bladder and urethra. Close the bladder with 5–0 or 6–0 monofilament, absorbable synthetic suture in a continuous, one or two layer inverting pattern. The bladder wall is very thin and care should be taken not to place any sutures in the lumen. Lavage the abdomen and close it with 4–0 or 5–0 synthetic absorbable suture in a routine pattern. Use systemic antibiotics that are safe for the guinea pig for at least three weeks postsurgically.
c Most urinary tract calculi in guinea pigs are composed primarily of calcium carbonate. Postsurgical management should be aimed at reducing the calcium in the diet, ensuring adequate fluid intake and treating any underlying bacterial infections. Recommend removal of alfalfa hay and limiting guinea pig pellets, both of which are high in digestible calcium. Use fresh grass hay instead of alfalfa hay. Use half a cup daily of a high vitamin C food, such as kale, mustard green, green peppers or cabbage for vitamin C supplementation. Stop the use of vitamin C in the water which may cause the patient to drink less due to a bitter taste. Some guinea pigs will drink more water if it is flavored with fruit juice, such as apple, cherry, pineapple or grape. Use an antibiotic that is safe for guinea pigs and that is consistent with the culture and sensitivity results. If no culture was taken or there was no growth, use an antibiotic, such as trimethoprim/sulfadiazine (30 mg/kg PO q 12 hours) or enrofloxacin (10 mg/kg PO q 12 hours) for a minimum of three weeks postsurgically.
d Even with this therapy some guinea pigs will develop urinary tract calculi again, therefore the prognosis should be guarded in cases of urinary tract calculi. Recommend frequent rechecks with complete urinalysis. Perform periodic abdominal radiographs.

210 a Does spend little time with their young so nests often appear deserted. They nurse as little as five minutes a day divided between two feedings, usually in the early morning and early evening hours. Recommend that the young rabbits be returned immediately to the nest if it has been 24 hours or less since they were removed. They will not be rejected by their mother due to the odor of humans. Nests can even be reconstructed without danger of rejection.
b The doe's milk is very high in fat and solids which allows infrequent feedings. Lamb milk replacer or canine milk substitute is mixed double strength or supplemented with cream. Even though a doe only nurses twice daily, it is difficult to reproduce the fat and calorie concentration of the milk, necessitating more frequent feedings for orphans. Feed with an eyedropper or feeding needle every six hours until the eyes open. After the eyes open, feed every eight hours. Stimulate urination and defecation by stroking the perineal area with a moist cotton ball.
c At 10–14 days of age.

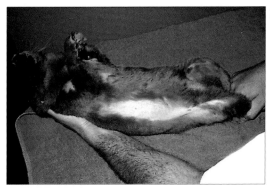

211 This two-year-old female rabbit is bred and 16 days later develops a ventral alopecia (**211**). The rabbit is fed a commercial rabbit pellet along with grass hay and fresh greens. There are no cagemates and it is allowed exercise outside the cage for one hour daily. The skin appears grossly normal.
a What is the most likely cause of the alopecia?
b How would you make a diagnosis?
c How would you manage the condition?
d What is the reproductive cycle of the female rabbit?

212 A ferret has been diagnosed with adrenal gland disease.
a What treatment would you recommend?
b If the owner elects not to treat this disease, what potential health complications could occur?

211 a Pseudopregnancy or pregnancy. Both pregnant and pseudopregnant rabbits will pull fur from their ventrum to construct a nest. The hair is pulled out intact and is not chewed or frayed. Other differential diagnoses to consider for alopecia in the rabbit include barbaring (either self or by cagemate), malnutrition, ectoparasites, dermatophytosis, mechanical due to rubbing, or endocrine.
b By abdominal palpation of the fetuses in the caudal ventral abdomen. In pseudopregnancy, the uterus may be slightly enlarged. Other signs of pregnancy or pseudopregnancy that may be seen include mammary gland and nipple development, increased aggression, nest building, hoarding of 'toys' and territorial marking with urine. The rabbit can be rechecked in a week with a radiograph to determine if fetuses are present. Pseudopregnancies may last as long as a pregnancy. Perform other diagnostic tests such as skin scraping, and microbiological culture as needed.
c If the rabbit is pregnant, hair is normally pulled out to make a nest therefore no treatment is necessary. It does not appear to be helpful to administer hormonal therapy to shorten the pseudopregnancy period. This is a benign condition that will resolve without therapy. Perform an ovariohysterectomy on females that are not to be used for breeding as a permenant cure for repeated pseudopregnancies.
d Does are induced ovulators with no definable estrous cycle. Ova are released within 13 hours of copulation. Periods of receptivity of up to 10 days are controlled by various factors, such as the photoperiod, ambient temperature and calorie intake. There are one to two day periods of inactivity when new ovarian follicles replace the old. Does can be receptive during pregnancy and lactation. There are a variety of signs of receptivity in the doe, but none are completely reliable in all rabbits. These signs include hyperemia and swelling of the vulva, increased territorial marking by chin rubbing or urination and a lordosis response when touched over the rump.

212 a Abdominal exploratory surgery to remove the diseased adrenal gland (see question **191**).
  Medical treatment is equivocal. Currently, recommendations are to administer mitotane for medical treatment. Mitotane decreases the signs of disease in some ferrets but its effects are usually not permanent. Use this drug with great caution in ferrets with insulinoma. Mitotane severely decreases the endogenous cortisol production such that ferrets with insulinoma can go into a serious hypoglycemic state and lapse into a coma. Therefore, screen all ferrets for evidence of insulinoma before instituting mitotane treatment. Use mitotane at 50 mg/kg PO q 24 hours for 7 days as an induction dose, and then q 2–3 days as a maintenance dose.
b At the least, the ferret may be bald and if it is a female it may have an enlarged vulva. The alopecia and enlarged vulva may resolve and recur spontaneously over time. More serious complications include estrogen-induced bone marrow suppression leading to anemia and pancytopenia. This is a rare complication of prolonged adrenal gland disease in ferrets but is life threatening. Adrenal gland disease leads to intense pruritus, although not a health complication, it is a side-effect that is difficult to control in a ferret. A large adrenal gland, especially right sided, can increase in size and put pressure on the vena cava. Finally, in male ferrets, a syndrome is recognized where the prostatic tissue around the urethra near the neck of the bladder increases in size under the influence of the androgens produced by the diseased adrenal gland.

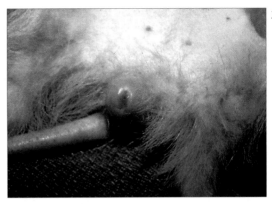

213

213 A four-year-old spayed female ferret progressively loses hair over its dorsum. On physical examination, the structure shown is observed (213).
a What is the structure?
b Why is this significant in a spayed female ferret?
c What are your differential diagnoses and what tests would you perform to develop a diagnosis? What is the most likely diagnosis?

214 Inhalant anesthesia is commonly used in rabbits.
a Which inhalant anesthetics are used most commonly in the rabbit?
b How would you implement inhalant anesthesia in the rabbit?

213 a A swollen vulva.
b A spayed female ferret should not have a swollen vulva. A swollen vulva is a natural occurrence in intact female ferrets. It is an outward sign of estrus. A swollen vulva in a spayed female ferret indicates she is not spayed, had an incomplete spay performed, or has endocrine disease present.
c Intact genital tract, ovarian remnants or adrenal gland disease. The differential diagnoses for alopecia include adrenal gland disease, seasonal hair loss, ectoparasites, other endocrine diseases and dermatophytosis. Alopecia and a swollen vulva in a ferret over two years of age are most likely due to adrenal gland disease. An abdominal ultrasound may detect an adrenal gland enlargement but frequently shows no change. Radiographs are usually not useful in this regard. Screening blood tests do not diagnosis this disease. The results of ACTH stimulation and dexamethasone suppression tests are normal since this is not Cushing's disease and therefore concentrations of cortisol are rarely elevated. Commonly, an increase in plasma adrenal androgens is present with this disease. The best way to diagnose this disease is to measure androgens such as androstenedione, DHEAS and 17-OH progesterone. Exploratory surgery may be the only way to make a definitive diagnosis.

214 a Isoflurane or halothane. Advantages of isoflurane anesthesia over the use of injectable anesthetics include the ability to rapidly change and control the plane of anesthesia and rapid recovery. Isoflurane is generally less of a cardiovascular depressant, has a more rapid onset of action and a more rapid elimination time than halothane. Therefore, isoflurane is preferred over halothane. Because it is less potent than halothane, a slightly higher concentration of isoflurane is required to reach a given plane of anesthesia. Inhalant anesthesia disadvantages include the cost of the agent and the requirement for an anesthetic machine, delivery system and careful scavenging of waste anesthetic gases. An endotracheal tube allows for control of the airway, facilitates ventilatory support, reduces environmental contamination with the inhalant and is preferred over a mask.
b Before inhalant anesthetic induction, animals can be premedicated with a tranquilizer such as acepromazine (0.1–0.5 mg/kg SC). For gas anesthetic induction, place the rabbit in a flow-through anesthetic chamber or, in depressed, sedated or easily restrained animals, use a mask and non-rebreathing circuit. Many rabbits tolerate chamber inductions well, even without tranquilization. Chamber induction without tranquilization allows only a short time for placement of the endotracheal tube before the rabbit wakes up. In non-premedicated rabbits, induction may require halothane or isoflurane at a concentration of 3–4%. Watch animals closely during chamber induction to prevent anesthetic overdosage. Minimize the release of anesthetic gases into the environment when using an induction chamber.
Alternatively, immobilize healthy animals with a combination of either ketamine and diazepam or ketamine and xylazine. Deepen anesthesia as needed by controlling the concentration of the inhalant. Isoflurane or halothane concentrations usually range from 0.5–2.0% depending upon premedications, the condition of the animal and the stimulus applied. If an endotracheal tube is in place, a small animal ventilator may be used to support ventilation. Minimize system leaks and properly scavenge waste anesthetic gases.

215

215 This adult male prairie dog exhibits a chronic non-pruritic generalized alopecia (215). The skin is lichenified with the presence of a few pustules. It lives with two others in a large dog cage with hay as the bedding material. The two cagemates have similar skin lesions but are less severely affected.
a What are the most likely causes of this problem?
b What diagnostic tests would you carry out and how is this condition treated?
c What is suitable housing and diet for a pet prairie dog?

216 The dietary requirements of the captive African hedgehog are still poorly understood.
a What common diet-related diseases are seen in pet African hedgehogs?
b What is an appropriate diet for a pet African hedgehog at our present level of knowledge?

215 a Parasitic or fungal dermatitis with secondary bacterial infection is the most common cause of these lesions in prairie dogs.

b Perform skin scrapings, skin biopsies and bacterial and fungal cultures. In this case, skin scrapings are negative for parasites. Fungal and bacterial cultures reveal heavy growths of *Trichophyton mentagrophytes* and *Staphlococcus aureus*. These animals were treated successfully with amoxicillin (10 mg/kg PO q 12 hours) for seven days and griseofulvin (25–50 mg/kg PO q 12 hours) for 10–14 days. The signs were completely resolved in four weeks. The hay may have been the source of the fungi in this case. A different bedding material was selected and the problem did not recur. Prairie dogs are hind gut fermenters with a GI physiology similar to that of guinea pigs. Although amoxicillin was used successfully in this case, use caution when selecting antibiotics for this species. There are no clinical studies currently available that show the safety or efficacy of antibiotics in prairie dogs. Use guidelines for antibiotic administration as set forth for rabbits. Other appropriate antibiotic choices would include enrofloxacin or trimethoprim/sulfa preparations.

c Prairie dogs are ground dwelling rodents native to North America. In the wild, prairie dogs live in large social groups within extensive 'towns' consisting of miles of interconnecting underground burrows. Captive prairie dogs need a large space for exercise and an area to satisfy their desire to dig. Provide a large sandbox at least 1.2 m deep for supervised exercise. Protect the bottom and sides of outdoor enclosures with heavy gauge wire mesh or concrete to prevent escape. Provide an indoor cage at least 1.2 m^2 with a thick substrate of pelleted bedding, aspen wood shavings or shredded newspaper. A nestbox filled with hay helps to satisfy the urge to dig and hide in a dark burrow. Use a metal cage because prairie dogs will quickly chew through wood.

An appropriate diet for a captive prairie dog consists of unlimited good-quality grass hay, along with dark leafy greens, small amounts of fruit and other vegetables. Feeding high calorie treats such as nuts, seeds, grains, sugars and dog or cat food can result in obesity.

216 a Obesity is the most commonly observed diet-related disease. It can result in secondary problems, such as lowered fertility rates and poor survivability of neonates. Other diet-related illnesses include hepatic lipidosis, dental disease, poor spine and hair condition, heavy scaling of the skin, nutritional secondary hyperparathyroidism and hypervitaminosis A and D.

b In the wild, hedgehogs feed on a wide variety of insects, small animals, fruits and vegetables. In captivity, maintain an average adult hedgehog on a daily diet consisting of 25 g of dry reduced-calorie cat food or mixture of dry and canned food, and 10 g of mixed fresh or thawed frozen vegetables. Three to four times per week, add 5 g (3–5 insects) of live crickets or mealworms. Since hedgehogs can easily become overweight, it is important to monitor the animal's weight frequently and adjust the amount fed appropriately. Feeding light or reduced-calorie cat food helps to maintain the animal's weight without compromising the nutritional quality of the diet. Live insects provide variety and behavioral enrichment for hedgehogs. Exercise is vitally important for the hedgehog species kept in captivity. Provide a large area in which the pet can exercise. Alternatively, use a large exercise wheel with a piece of wire or plastic mesh fitted on to the inner surface to prevent the feet from slipping through and to provide traction.

Classification of cases

Abbreviations

ACTH   Adrenocorticotropic hormone
BUN   Blood urea nitrogen
CAR   Cilia associated respiratory
CBC   Complete blood count
CNS   Central nervous system
CT   Computerized tomography
ELISA   Enzyme-linked immunosorbent
    assay
Fr   French
GI   Gastrointestinal
GnRH   Gonadotropin releasing hormone
HCG   Human chorionic gondaotropin
ID   Internal diameter
IFA   Immunoflourescent antibody
IM   Intramuscular

IO   Intraosseous
IP   Intraperitoneal
iu   International unit
IV   Intravenous
MCV   Mean corpuscular volume
MRI   Magnetic resonance imaging
PCV   Packed cell volume
PO   Per os (by mouth)
q   Every
RBC   Red blood cell
URI   Upper respiratory infection
SDAV   Sialodacryoadenitis virus
SC   Subcutaneous
WBC   White blood cell

Index